Concrete Revolution

Concrete Revolution

Large Dams, Cold War Geopolitics, and the
US Bureau of Reclamation

CHRISTOPHER SNEDDON

The University of Chicago Press
Chicago and London

Christopher Sneddon is associate professor of geography and environmental studies at Dartmouth College.

The University of Chicago Press, Chicago 60637
The University of Chicago Press, Ltd., London
© 2015 by The University of Chicago
All rights reserved. Published 2015.
Printed in the United States of America

24 23 22 21 20 19 18 17 16 15   1 2 3 4 5

ISBN-13: 978-0-226-28431-6 (cloth)
ISBN-13: 978-0-226-28445-3 (e-book)
DOI: 10.7208/chicago/9780226284453.001.0001

Library of Congress Cataloging-in-Publication Data
Sneddon, Christopher, author.
    Concrete revolution: large dams, Cold War geopolitics, and the US Bureau of Reclamation / Christopher Sneddon.
        pages: illustrations; cm
    Includes bibliographical references and index.
    ISBN 978-0-226-28431-6 (cloth: alk. paper) — ISBN 978-0-226-28445-3 (e-book) 1. Dams—History—20th century. 2. Water resources development—United States. 3. United States Bureau of Reclamation. 4. Water resources development—Political aspects. 5. Technical assistance, American—Political aspects—Developing countries. 6. Geopolitics—United States. I. Title.
    TC556.S63 2015
    333.9100973—dc23   2015002173

♾ This paper meets the requirements of ANSI/NISO Z39.48-1992 (Permanence of Paper).

CONTENTS

75/90 #

ILLUSTRATIONS

ACKNOWLEDGMENTS

This book would have been inconceivable without the support and encouragement of friends, family, and colleagues over the past two decades. It has been an arduous process at times, and I am deeply thankful for their support and encouragement. Their ranks are too numerous to identify everyone by name, but I do want to draw attention to those people who have been instrumental in guiding me (on occasion unwillingly!) down a path that kept redirecting me toward understanding the phenomenon of large dams. While many did not offer overt commentary on the contents of this work, they nevertheless inspired my thoughts on dams, rivers, geopolitics, and history in ways that certainly inspired the ideas and narratives set down on paper.

In the early 1990s I had the great fortune to work with two extraordinary thinkers and people at the University of Michigan. The late Bill Stapp first suggested the Mekong River basin as a potential paper topic during a seminar, and Richard Tucker taught me how to think like an environmental historian and to consider the connections between dams and geopolitics. I am still following his perceptive advice. My time later in the decade at the University of Minnesota was a period of great intellectual creativity and numerous conversations about dams. In particular, I want to thank Leila Harris, Jim Glassman, Tsegaye Nega, Abdi Samatar, and Eric Sheppard. Allen Isaacman and all of the wonderful people affiliated with the MacArthur Program (now the Interdisciplinary Center for the Study of Global Change) in Minnesota offered intellectual vibrancy and wonderful friendship in equal measure. At various times, two giants of interdisciplinary environmental scholarship—Richard Norgaard and the late Fred Buttel—reinforced in me a belief that I had something to contribute. I have also drawn numerous insights from the textured, thoughtful work of Rod Neumann, whose ways of thinking about human-environment relations inspire my own.

The archival research for this project was generously facilitated by the assistance of Brit Storey, Roy Wingate, and Marene Baker at the NARA Rocky Mountain facilities in Denver, and Eugene Morris played a similar role in NARA's College Park archives. In Thailand and at visits to the Mekong River Commission in Vientiane, Lao PDR, I was greatly helped by Apichai Sunchindah, Vitoon Viriyasakultorn, and Somsak Wichean. Wolf Hartmann has been a kind and considerate presence in discussing all things Mekong-related. The majority of the research undertaken to produce this volume was generously supported by the National Science Foundation under Award #0823197. I also received financial support from the Dean of Faculty, Dartmouth College, to assist in research visits. Portions of chapter 3 were published previously in *Political Geography* (doi:10.1016/j.polgeo.2011.09.005) and segments of chapter 5 in *Social Studies of Science* (doi:10.1177/0306312712445835). I am grateful to Elsevier and Sage Journals, respectively, for permission to reuse this material.

During its gestation over the past five years, I had the wonderful opportunity to present parts of the book at Macalester College, the University of Minnesota, the Massachusetts Institute of Technology, the University of Wisconsin, and McGill University and want to thank the audiences in those venues for lively and thoughtful comments. Gratitude is due to all my colleagues at Dartmouth, but I want to single out Frank Magilligan, Mona Domosh, Richard Howarth, Richard Wright, and Anne Kapuscinski (among others) for their willingness to listen to complaints and make timely suggestions, not to mention their general collegiality. My oftentimes coauthor Coleen Fox has been a constant source of motivation in thinking about rivers, dams, and politics. Jonathan Chipman has my sincere gratitude for his superb cartographic skills, demonstrated by the wonderful original maps exhibited throughout this book. A special thanks is due Samer Alatout, whose friendship and intellect are present in these pages. I also owe a great deal to the reassuring advice and comments from Christopher Chung, Abby Collier, and Christie Henry at the University of Chicago Press. I deeply appreciate the thoughtful suggestions of two anonymous reviewers of the original manuscript; their ideas have greatly enhanced the final outcome. This book would have been far more arduous without the almost daily injections of humor from the baristas and fellow patrons of the Tuckerbox Café in White River Junction, Vermont.

Finally, none of my scholarly activities would be possible without the love and support of my family. My parents, Patricia and Boyd Sneddon, and sister, Cindy Sneddon, have been wonderful sources of encouragement. My amazing partner, Connie Reimer, and our two wonderful children, Maeve and Ethan, always remind me of the most important things in life and work. As the saying goes, all errant statements in the book are my sole responsibility.

ABBREVIATIONS

AHC	American Heritage Center
AUFS	American Universities Field Staff
DAC	Development Assistance Committee
DDRS	Declassified Documents Reference System
DRC	Development Resource Corporation
DSI	Devlet Su Isleri (Turkey)
ECAFE	Economic Commission for Asia and the Far East
FOA	Foreign Operations Administration
FRC	Federal Records Center
FRUS	Foreign Relations of the United States
GERD	Grand Ethiopian Renaissance Dam
GPO	Government Printing Office
IBRD	International Bank for Reconstruction and Development
ICA	International Cooperation Administration
ICOLD	International Commission on Large Dams
IDB	Inter-American Development Bank
IEC	International Engineering Company
IEG	imperial Ethiopian government
IWRM	integrated water resource management
LMI	Lower Mekong Initiative
LRA	Litani River Authority
NARA	National Archives and Research Administration

NRC   National Resources Commission

NSC   National Security Council

OECD   Organisation for Economic Co-operation and Development

RG   Record Group

STS   science and technology studies

TVA   Tennessee Valley Authority

UNESCO   United Nations Economic, Social and Cultural Organization

USAID   United States Agency for International Development

USG   United States government

USOM   United States Operations Mission

WCD   World Commission on Dams

ONE

Large Dams, Technopolitics, and Development

Historians and other observers have demarcated the twentieth century according to a wide variety of cogent social and biophysical features. Some of the most popular candidates for "century-defining" trends include urbanization, rapid population growth, agricultural production, and more ominously, its characterization as the "most murderous century of which we have record."[1] Yet the twentieth century also witnessed a radical transformation of the planet's river systems through the construction of an estimated 50,000 large dams, a hydrological and ecological experiment that has fundamentally altered human relations with water.[2] One of the central ambitions of this book is to uncover the specific ways in which large-dam technologies and the ideologies that guided them have proliferated across the planet in the twentieth century. These ideologies and technologies are deeply intertwined and serve a central role in explaining how this "concrete revolution" materialized in the relatively short span of seven decades. My particular focus falls on the activities of the United States government to promote and shape the dissemination of, first, a crucial technological innovation in the form of large-scale hydroelectric dams and, second, a novel approach to resource use in the form of river basin planning and development. Throughout the Cold War era, these activities were largely carried out under the auspices of the United States' preeminent water resource development agency, the Bureau of Reclamation, and were in many cases directed by the geopolitical imperatives of the State Department, who saw technical assistance as a crucial tool in staving off the presumed global expansion of communism. Technical acumen and geopolitical imagination came together in a methodical process of damming the planet.

Large dams are perhaps the quintessential example of what scholars across a range of disciplines call nature-society hybrids. These massive structures, whether formed of concrete, or earth and rock, or more likely some combination, reside at the intersection of complex networks of altered hydrologies, technical expertise, financial circuits, political desires, displaced communities, and hegemonic ideologies. Dams, particularly since the 1970s, have also been the focal point of intense social conflict. The publication of *Dams and Development: A New Framework for Decision-Making,* a seminal 2000 report by the World Commission on Dams (WCD), was the culmination of over three decades of debate on the benefits and costs of large dams.[3] In brief, the report concludes that although dams "have made an important and significant contribution to human development," they have too often produced severe social and environmental impacts borne disproportionately "by people displaced, by communities downstream, by taxpayers and by the natural environment."[4] Predictably, the report was greeted with contradictory responses upon its release. Representatives of the global dam-building industry and government representatives of countries with active dam-building programs, such as China and India, condemned the report as disingenuous, lacking in rigor, and irrevocably biased against large dams.[5] In contrast, the global anti-dam movement fully endorsed the WCD report and its guidelines and called for immediate action on the part of governments and international financial institutions to implement its recommendations regarding more participatory and transparent governance of water resource development. Despite the scads of information in the report regarding the world's dams and the varied reactions to its conclusions, both data and responses were remarkably ahistorical, shedding little light on the practices and negotiations that over the course of the twentieth century brought forth so many thousands of large dams. Remarkably, perhaps trying to appear balanced as an international body seeking common ground within a highly charged debate, the WCD made little mention of the political character of large-dam projects.[6] The study proposed here argues, conversely, that the construction of large dams and the ideas set forth under the rubric of river basin planning, as well as the socioecological transformations wrought by these activities, are inseparable from the political dynamics among the social actors who mobilized and sustained these technologies and ideas in the first place. Dams are, as a matter of course, exceptionally "thick" with politics.[7]

It was the geopolitical thickness of large dams and associated river basin development schemes that promulgated a concrete revolution in the twentieth century and hence prompted the title of this book. Although this phrase is designed to mirror the other "revolutionary" developments of the same period—the Green Revolution being the obvious referent—I do not use it glibly. Large dams, brought into being through a combination of technological prowess, engineering expertise, and political-economic calculation, have radically altered humanity's relationship with planetary river systems. What word other than "revolutionary" would one use to describe a human intervention into socioecological processes that has directly displaced somewhere between 30 and 60 million people,[8] deleteriously affected the livelihoods of roughly another 500 million people situated downstream,[9] and converted tens of thousands of riverine environments into highly regulated water systems at a scale unprecedented in the history of the planet?[10] This global transformation has been concrete in the obvious sense of the pouring of countless tons of cement, water, sand, and gravel into forms that engineering designs and human labor shape into dams.[11] Yet this concreteness is also found in the physicality and durability of dams. Indeed, it is hard to imagine a more massive or widespread technological intervention that, in its profound and active materiality, has so challenged, if not erased, the boundaries between technology, humanity, and nature. The transformative aspect of this concrete revolution resided not only in its capacity to provide millions of kilowatts of electricity "for the lamps of China," according to one effusive journalist writing of plans to dam the Yangtze River in the mid-1940s.[12] Like its affiliate, the Green Revolution, the concrete revolution was deeply implicated in global geopolitics and efforts by the United States foreign policy apparatus to exert influence over newly emerging nation-states via technical and economic assistance.[13] Large dams were both forged in and helped congeal a revolutionary agglomeration of water and geopolitics.

My argument is thus built around two central, interrelated themes concerning the proliferation of large dams and river basin development spanning the period, roughly, from 1933 to 1975, coinciding approximately with the era of Cold War frictions between the United States and the Soviet Union. The first theme concerns the deep linkages among geopolitics, technologies, and large-scale environmental transformations carried out in the name of "development." Large dams and other so-called megaprojects were intimately connected to ideas of modernization and hence, in the context of US and Soviet outreach to decolonizing regions of Asia and Africa following World War II, became a preferred form of economic development assistance during the height of the Cold War.[14] Engineers and development planners alike expected large dams to revolutionize economies and societies through electricity production and irrigation development. In contrast, for the architects of post–World War II foreign policy in the United States, the transfer of technological expertise regarding water resource development was a crucial way of solidifying geopolitical alliances between the American state and a host of newly independent postcolonial regimes in Asia, the Middle East, and Africa.[15]

A second theme centers around the production and transfer of a powerful geographical ideal: that of the river *basin* as the most appropriate unit for a host of interrelated water development and management activities. This notion is now firmly rooted within the planning agendas of an array of state agencies, international financial organizations, and nongovernmental organizations. However, visions of the basin as an integrated developmental unit did not emerge from the ether.[16] Throughout the twentieth century, technological proficiency (some might say hubris) and geopolitical objectives combined to produce a potent image of basins as the primary vehicle for developing the potential of rivers, and of dams as the key technological vehicle for achieving this dream. Moreover, this image of the basin's holism rested firmly on the capacity of dams to confer control over water to human managers and the governments they serve. Large dams offered the material capacity to profoundly alter rivers, while river basin development provided the institutional and managerial scaffolding. Ultimately, the geopolitical architecture of the Cold War provided a nearly perfect political environment for the rapid spread of large dams and associated ideas of the river basin.

These geopolitical, technological, and developmental processes coalesced in the work of the United States Bureau of Reclamation, the United States', and perhaps the world's, premier dam-building bureaucracy of the twentieth century. The Bureau, an agency of the US Department of the Interior responsible for designing and constructing many of the large-scale water development projects in the western United States during the twentieth century, constituted a small but highly influential element of the proliferation of large dams throughout the Cold War period. In response to increasing requests for technical assistance from "underdeveloped" countries, the Bureau's international activities emerged in the later 1930s, blossomed in the 1950s, and continued to grow in the 1960s, eventually including active missions in over 50 different nation-states and providing some form of technical assistance to well over 100 countries. During this period, the Bureau's activities in international affairs encompassed technical services (including "review and analysis of designs, special engineering studies, performance of laboratory tests, and preparation of reports"), technical missions consisting of individual engineers or groups of engineers working as advisers to host nation personnel, training programs for foreign engineers over intervals ranging from twelve months to a few days (for observations by more highly skilled engineers), dissemination of technical publications, and participation in overseas conferences, workshops, and other forums.[17] These technical missions are this book's primary focus. The goals of these assignments almost always included on-the-ground assistance in determining the feasibility of large dams and the potential for river basin development; the training of host foreign nationals in the multifaceted nature of dam construction and river basin planning; and aid in the formation of appropriate water bureaucracies. These activities were channeled through the US State Department and the national security advisers of successive US presidents, from Truman and Eisenhower in the 1950s to Kennedy, Johnson, and Nixon in later decades. In nearly all instances, the technical expertise of the Bureau's staff members came into conflict with the geopolitical agendas of the architects of American foreign policy. The chronicle of large dams could and should be related through a variety of institutional lenses, but the Bureau's pivotal role as purveyor of technical expertise and translator of geopolitical imaginations offers a unique point of departure for examining the complex technopolitical networks of water development that arose in the twentieth century and persist to this day.

What were the geopolitical rationales that guided this profusion of technical interventions? How should we understand the interrelations among technology, environment, and politics that both facilitated and thwarted the diffusion of dams? The rest of this chapter outlines a conceptual framework for querying the profusion of large dams and river basin ideology throughout the tricontinental world from the mid-1930s until the mid-1970s.[18] Large dams—as technological objects constituted through assemblages of capital, knowledge, and power—represent a crucial spatial and temporal node of technopolitics in the twentieth century. The moment has surely come to stop thinking about large dams, and indeed, all technology-centered development projects, as purely technical undertakings whose successes or failures hinge on the ingenuity of the engineers who design and build them or on the motivations of state officials who fund and promote them. As an inspection of history shows, dams and the processes and things they draw together have never *acted* in this way. Rather, the lessons of the history presented here are that large dams and river basin planning are complex hybrids of nature, technology, and society. These hybrids behave in often unpredictable ways, despite the best efforts to plan for and take account of the social and biophysical changes wrought by damming a river. Indeed, for most of the twentieth century, the socioecological transformations produced by large dams were mere afterthoughts to their geopolitical and developmental utility.

Large Dams, Technopolitics, and
the Hidden Legacies of the Cold War

Why do we live on a dammed planet? Conventional responses have built their answers around a few straightforward propositions. For proponents of hydroelectricity production and the expansion of irrigated agriculture, large dams are an elegant technological solution to global society's ever-increasing demand for industrial development and expanded food production. The impoundments they create represent a culmination of the efforts of human ingenuity to exert control over the vagaries of natural processes. Their proponents argue that unruly rivers can and should be harnessed for the sake of human betterment. Key to this understanding has been the assumption that moving water will benefit humanity most efficiently by being converted to electricity or stored for irrigation, flood control, navigation, human consumption, and other potential benefits. Critics counter that dams are a product of the profound hubris of technological optimists and their supporters in government and industry, who have stoppered river systems in ignorance of the potent biophysical impacts of damming and with carelessness toward the substantial human costs of displacement and loss of livelihood.[19]

As powerful as both these narratives—dams as technological saviors and dams as destroyers of rivers and riverine people—have been, both to my mind have insufficiently grappled with the question of *how* so many dams came into being over the course of the twentieth century. In other words, they have ignored the genealogy of large dams and, concomitantly, that of river basin development.[20] Thus what we have learned from historical studies of large dams and river basin development—that the development of water is deeply tied to the expansion of capitalist agriculture, is often bound up with regional economic evolution, and is part and parcel of nationalist and modernist agendas—is valuable and useful, but still leaves a crucial component of the genealogical question unexamined. While I use different words in this book to describe the diffusion of large dams across the face of the planet, perhaps none is more apt than "proliferation." I suggest that large dams, particularly the hydroelectric dams that emerged in various parts of Asia, Africa, the Middle East, and Latin America during the Cold War, are indeed lively things. More than any of their other legitimating rationales, what came to be called hydropower and its insertion into burgeoning electricity transport systems (materially and emblematically "networks of power") propelled the rapid spread of large dams.[21] These dams have rewritten the face of the planet, inscribed a new set of biophysical relations within the river basins where they were constructed, and thus represent a novel geo-*graphy* that is planetary in scope. A corollary question to why nearly all the planet's major rivers are dammed thus becomes, how did the geopolitical visions adopted by the US state establish the groundwork for novel geographies of development via the diffusion of large dams and the technical expertise that accompanied them? A primary goal of this book, then, is simply empirical: to detail a significant current of the history of damming and its subsequent impacts.[22]

Dams have certain characteristics that make them a central actor in the history of development practice over the course of the past century. I use "actor" not only metaphorically, but also as it is used in actor-network methodologies, where things act in the sense of having effects on a host of human and nonhuman processes that are independent of their creators' intentions and designs.[23] Large dams act as central hubs that draw together—or assemble—various kinds of networks. One of the advantages of conceptualizing dams in this fashion is the shift it demands in the unit of analysis. No longer are the dam, its reservoir, and the socioecological alterations it visits on a particular river system a sufficient explanatory framing. If the goal is to understand the origins of large dams and how they are situated within broader constellations of nature-society relations, there must be an accounting of the variety of networks (e.g., financial, symbolic, ecohydrological, and so on) through which the actualization of a specific dam occurs. My particular focus lies in the particular region of the "dam assemblage," where technoscientific networks of engineering expertise intersect with geopolitical dynamics and their historical trajectories. This is precisely where the Bureau of Reclamation intervened and subsequently influenced the global proliferation of large dams.

My adoption of the terms "assemblage" and "technopolitical network"—which have emerged from recent work in science and technology studies (STS) and related fields—raises challenging epistemological and methodological issues regarding large dams and their capacity to bring together and maintain associations of human and nonhuman agents. I use "assemblage"—a term that has arisen within several strains of social theory—to denote the collection of things, places, and processes brought together by a central idea or material entity.[24] In the abstract, assemblages are simply "ensembles of heterogeneous elements,"[25] and these elements themselves "may be human and non-human, organic and inorganic, technical and natural."[26] Large dams can be fruitfully thought of as assemblages because they bind together the hydrological and ecological processes of large river systems, flows of capital, economic development ideals, geopolitical agents, and (crucially) the technical expertise of human engineers (for example, those in the Bureau of Reclamation). Importantly, the notion of assemblage emphasizes the conditional character of things and processes brought together under a common rubric such as "large dam" and the oftentimes unexpected outcomes that arise from such collectives.

While I say more about technopolitics and technopolitical networks later in the chapter, I see "assemblage" and "network" as highly complementary metaphors for describing the complex interrelations among people, technologies, and ecologies that emerge at particular historical junctures.[27] In an effort to overcome the numerous dualisms that have characterized much of social theory in the twentieth century (e.g., human/nonhuman, macro/micro, object/subject, active/passive, specific/general), both metaphors accentuate the need to shift our units and scales of analysis to account for the complex webbing that joins relations and things. As mentioned previously, they also shift to whom and to what we ascribe agency, allowing the possibility that active participation in the world—while not necessarily exhibiting intentionality—"arises from collective endeavor," and that one therefore needs to exhibit an "ecological" mind-set in order to fully appreciate the range of human and nonhuman "bits and pieces" whose collective relations produce all kinds of effects.[28]

An important contribution to these non-dualistic world views is provided by recent work in geography, which sees the idea of assemblage as especially useful "to stress the *making* of socionatures whose intricate geographies form tangled webs of different length, density and duration, and whose consequences are experienced differently in different places."[29] In my understanding, assemblages woven together around large dams create new geographical (spatial) relationships between sites of construction (e.g., in Ethiopia or Lebanon), broader development plans centered on the river basin, and global geopolitical forces such as those associated with the Cold War. "Assemblage" is thus an apt description for the types of geopolitical relationships, technological knowledges and practices, and biophysical dynamics brought together in multiple places around the globe as a result of the globalization of large dams and river basin planning during the Cold War. Theorizations of assemblages, or collectivities, of human and nonhuman agents direct attention away from a somewhat vague conceptualization of the environment as "socially constructed" and toward the multiple and convoluted processes—be they the Cold War geopolitical strategies of an imperial state, the technological challenge of building a large hydroelectric dam in the tropics, the struggles of postcolonial states to simultaneously promote economic development and national identities, or the ecohydrological dynamics of large river systems—through which particular "natures" (those signified by altered river basins) are constructed and contested.[30]

Geographers and others have also nudged assemblage and network approaches toward a greater appreciation of questions regarding the exercise of power, or who and what has the capacity to assemble and sustain multiple humans, technologies, ideas, things, and biophysical processes in more or less dense networks of relations.[31] In order to understand these relations and extensions of power and the concrete effects they produce, there is a need to examine collective action in all its complexity, from how actors come together to how they sustain collaborative efforts.[32] Ultimately, the mobilization and diffusion of dam-building techniques and river basin planning approaches is impossible to understand without reference to power as a set of relationships, some more resilient and some working at broader scales than others. In the terms of assemblage thinking, power dynamics are what define the duration of any given assemblage, its relative value within society, and the human actors who might benefit from the actions of the assemblage at some point in its history.[33] Relatedly, the entities that constitute actor-networks—technopolitical or otherwise—do not arise in a vacuum; they are built and consolidated via processes over which some beings typically have more control than others.[34] It is this set of power relationships—involving but not limited to differentiated states, technical experts, construction firms, displaced peoples, altered rivers, and so on—that is inherent in massive biophysical alterations in the name of "economic development" and so often overlooked in contemporary debates about large dams and river basin development. Too often this debate, perhaps epitomized in the divergent responses to the WCD report, is reduced to a caricature, a simple choice of "the environment" and "rural livelihoods" versus "development" and "modernization." Both sides of this debate, I suggest, depend on a radical separation of the human and nonhuman and on an inattention to how power works over time and space.

Finally, conceiving of large dams as assemblages has important implications for how we understand our units of analysis in a broad array of research traditions within the social, human, and natural sciences. If large dams are nodes in a series of networks that bring together political-economic, technological, financial, ecological, hydrological, and cultural processes, our explications of the amalgamation and maintenance of specific network configurations are critical. Circulating throughout all these processes, and in some respects integrating them, is a specific sort of technopolitical knowledge that must be brought to light. I argue that this knowledge is best captured through a combination of thinking on the historical dimensions of dams and their socioecological transformations, the geopolitics of development, and finally, novel ways of capturing the complexity of nature-technology-society relations.[35] This book also asks what happens when technopolitical networks, forged under one set of institutional and environmental conditions, are set down and mobilized within quite different geographical and historical contexts. As subsequent chapters ask, what happens when dams travel? A response demands attention to the historical circumstances of their dissemination and to the imbrication of politics and technical expertise that propels their movements.

Historicizing River Alteration

At least two veins of historical inquiry are relevant to understanding the relatively rapid spread of large dams and basin-oriented approaches to water development throughout the tricontinental world in the Cold War era.[36] The first vein is best described as environmental history concentrated primarily on water and technology. Broadly, environmental history focuses on the shifting relationships between humans and the biophysical environment, as well as on the complex evolution of human interpretations of "nature" over time.[37] A growing literature has used water as a focal point for examining these histories, delineating the complex relationships among, for example, government strategies to promote the development of water resources, transformations of river systems and coupled landscapes, and associated responses in human social organizations to adapt to these novel hydrological conditions.[38] These theoretically ambitious, deftly articulated histories encourage an approach that casts a wide analytical net, combining elements of the institutional and political contexts through and around which human productive activities contributed to the transformation of river systems in very specific ways. This book is certainly an environmental history, but one with qualifications. While the material transformation of river systems remains a crucial historical concern, my central goal is better described as tracing the emergence of the ideology and technology of large dams—and the consequences of this emergence—somewhere in the middle ground between thing and idea. Moreover, subsequent chapters demonstrate that the environment—in the form of hydrological processes, geological conditions, alterations to fisheries, and many other biophysical processes—remains a vitally important element in the geopolitics of large dams.

Historians of technology-society relations have also exhibited a sustained engagement with the interactions between water and society. Research examining the manipulation of water resources has elucidated the evolution of particular technologies, histories of specific water resource development projects, and organizational histories of key water management bureaucracies, albeit almost entirely within a US context.[39] This scholarship has added much to our empirical knowledge of crucial historical periods in humankind's capacity to alter aquatic systems, but few studies have sought to explain the broader societal forces that generate and maintain technology-driven development of water resources. Moreover, histories of water resource use, seemingly a logical point of intersection for environmental history and the history of technology, have infrequently examined the technological and environmental agents of transformation in tandem.[40] In addition, water- and river-oriented environmental historians have eschewed more global investigations, focusing instead on particular projects, particular river basins, or particular geographical regions.[41] One of my central arguments is that to fully comprehend the social and biophysical changes brought about by large dams and the accompanying ideology of river basin development, one has to transcend political boundaries and explicate the evolution of the global network of dam building and river basin planning.[42]

Another key goal of this work is to draw out the links between, on one hand, the activities of an American bureaucracy representing technical expertise and, on the other, the broader geopolitical strategies of a succession of US regimes intent on containing the spread of Soviet Communism and securing the conditions for sustaining US global economic dominance. I thus draw additional inspiration from a second vein of historical work that constitutes a retelling of the Cold War from a "Third Worldist" and developmental perspective. Grouped loosely under the heading of "new historicism," scholars working within this rubric underscore the multiple ways in which geopolitical relations defined in terms of an East-West dichotomy shaped the political structures, economic relations, and social dynamics of societies in the "underdeveloped" regions and, in turn, how these structures, relations, and dynamics fed back into global geopolitical calculations shaped by the Cold War.[43] Building on the proposition that the Cold War did indeed have a global reach, I suggest that the diffusion of large dams and the concomitant spread of the discourse and practice of river basin development throughout the nation-states of the so-called Third World was a critical facet of efforts by the US State Department to use "economic development" as a bulwark against what US officials perceived as the global expansion of communism.[44] The basic idea was to demonstrate the benefits of capitalist economic development through the transfer of financial resources as "development assistance" and of "development expertise" in the form of novel technologies and planning approaches, with (from the perspective of the United States) the more implicit goal of protecting US business interests and potential overseas markets.[45] Indeed, an excavation of the intertwined historical, technological, and geopolitical networks driving the proliferation of large dams would be incomplete without reference to the political-economic forces also contributing to the interventions of postcolonial states in river questions. There is no question that the United States pursued its Cold War geopolitical agenda in tandem with the expansion of capitalist development.[46] The global expansion of large dams and river basin planning under the auspices of the Bureau of Reclamation is indicative of these trends, but remains a largely hidden element of the broad swath of research on the global Cold War.

Historicizing the connections between the spread of large dams and Cold War geopolitics is a critical aim of this book, but there is an equally important obligation to theorize these connections and demonstrate how this history might contribute to ongoing (and current) discussions regarding the geopolitics of development and nature-society relations.[47] The agency of the Bureau and its subsequent efforts to promote river basin development were animated by a particular geopolitical vision that found expression in US foreign policy in the years following World War II. This vision was firmly grounded in theories of economic development and modernization of the time.[48] The US government regarded the newly independent states of Asia, Africa, and other "underdeveloped" regions—if it regarded them at all—as economically backward and politically immature territories susceptible to the potent ideological influence of the Soviet Union and its anti-American stance.[49] Modernization theorists influential during this period expressed fears that "the USSR was providing a better example of development," and believed that the "Soviet threat [was] the essential starting point for thinking about development."[50] In the case of US efforts to promote river basin development as a Cold War "weapon," global political relations were spatialized—that is, divided up according to geographical categories—in such a way as to create a Third World ripe for technological intervention.[51]

Given such spatializations, work within the geopolitics of development is especially germane to geographical concerns with human-environment transformations, political-economic dynamics, and the historically specific discursive strategies of both the providers and recipients of "development assistance."[52] Applied to analyses of the history of river basin planning and the diffusion of large dams, approaches informed by critical geopolitics make it possible to clarify how geopolitical discourses actively construct the "river basin" as a fulcrum of development within a representational frame that privileges "national" territories and certain aspects of ecological networks. The basins that eventually became the foci of the Bureau of Reclamation's work were imagined by the State Department, and to some extent by Bureau planners, as discrete geographical entities whose development could be universalized and extended to virtually every nation-state in the world. The result of this geographical imagination of the basin is a considerable simplification of a complex political-ecological entity, one that, in the language of James Scott, is made legible and thus amenable to manipulation by state administrative apparatuses.[53] Critical geopolitics draws attention to the discourses that underpin such simplifications and how they are generated by centers of power to serve a given geopolitical order, revealing how both geographical knowledge and geographical representations shape understandings and practices of seemingly apolitical development initiatives. In the twentieth century, both technologies—large dams, for example—and certain ways of imagining the political geography of the planet became critical components of the expansion of American hegemony.

At the level of world politics, hegemony connotes a condition in international relations whereby one state is able to enroll "others in the exercise of [its] power by convincing, cajoling and coercing them that they should want what you want."[54] Robert Cox emphasizes that hegemonic power operating at a global scale in the sphere of international affairs is constituted by and reflective of hegemonies of social classes within particular nation-states. Hegemony in world politics is "in its beginnings an outward expansion of the internal (national) hegemony established by a dominant social class," and (eventually) the "economic and social institutions, the culture, the technology associated with this national hegemony become patterns for emulation abroad."[55] But hegemony in the global sense delineated by Agnew and Cox must be contemplated at other scalar levels as well. Accordingly, my focus on the Bureau of Reclamation and its institutional collaborators in various apparatuses of the State Department seeks to crack open the "black box" of the organizational structures actually charged with implementing policies designed to advance America's hegemonic aims.[56] The Bureau eventually became a key locus of the expression of American geopolitical power by virtue of its technical expertise in water resource development and its explicit identification as an agent of development and modernization. Seeing global power relations as fundamentally hegemonic adds a degree of suppleness to analysis of how seemingly "neutral" technological assistance and the promotion of large-scale water infrastructure are in fact entangled with states' geopolitical designs.

Large Dams and Technopolitical Networks

In tandem with historicizing the geopolitics of large dams, this book also addresses the growing scholarly interest in technopolitics. The architects of foreign policy within the US government initiated, guided, and sustained numerous Cold War–era efforts to apply technical and social scientific knowledge to the combined geopolitical and economic problems of the post–World War II era. This complex co-production of technology and politics can be termed "technopolitics." During this period, technologies also played critical roles in mediating the political and economic relations between powerful and weaker states and fomenting a host of socioecological trans-formations.[57] Large dams are exemplary in this regard. For example, Timothy Mitchell's analysis of the origins, construction, and outcomes of the projects at Aswan on the Nile River (both the first "low dam" built by the British colo-nial regime at the turn of the twentieth century—later raised in the 1930s—and the more familiar Aswan High Dam completed in 1971) revolves around the radical historical contingencies of Cold War geopolitical strategies; rural development programs and the responses of Egyptian farmers; the bio-physical dynamics of the Nile River itself; the host of unintended ecological changes wrought by the river development projects; and the machinations of Egyptian elites vying for political power during the twentieth century. Cir-culating throughout this array of social, economic, cultural, and biophysi-cal processes was a struggle over knowledge as well as the emergence of a technological expertise that posited, and continues to posit, a radical separa-tion between human and nonhuman, between modern and primitive, and between local and universal. Mitchell's understanding of technopolitics thus beckons toward the tightly linked, highly conditional, and unpredictable characteristics of relations among social actors, technologies, and ecological processes:

> Techno-politics is always a technical body, an alloy that must emerge from a process of manufacture whose ingredients are both human and nonhu-man, both intentional and not, and in which the intentional or the human is always somewhat overrun by the unintended. But it is a particular form of manufacturing, a certain way of organizing the amalgam of human and nonhuman, things and ideas, so that the human, the intellectual, the realm of intentions and ideas seems to come first and to control and organize the nonhuman.[58]

Technologies, to reiterate, serve as assemblages of networks, but if we follow Mitchell's lead, ones that are only partially coordinated by humans.

Despite a long-standing interest in technologies as expressions of political desires and conflicts, a growing and stimulating engagement with "tech-noscience" as a form of political negotiation, and broader commitments to examine the multiple and fluid interactions between technology and politics,[59] scholars of science and technology studies have had surprisingly little to say about dams as technological and political objects or about their origins in developmental impulses. This absence is rather remarkable considering the highly politicized origins and impacts of so many large dams, the vast societal expenditures required for their construction, their material and symbolic attachment to governments and even specific regimes, and their capacity to simultaneously and radically transform biophysical and social relations. It may, perhaps, relate to the sheer brute physicality of dams and the often extensive periods (years or even decades) between their conceptualization and actual construction. Perhaps even more notably, STS has infrequently engaged with geopolitics in discussions of the technology-politics nexus, particularly the historical geopolitical architectures of the Cold War.[60] Hence focusing on the origins and spread of large dams leads to a dialogue between thinking within STS and work in critical geopolitics and the geopolitics of development. Geopolitics, as noted above, encompasses the historical processes that produce state structures as well as the geopolitical imaginations, grounded in discourses specific to certain times and places, that give rise to different modes of geopolitical expression (e.g., formal, practical, and popular).[61] Turning toward the development vision, there comes a key point in nearly every development project when the array of complex political, ideological, cultural, and ecological processes that constitute any particular site are "rendered technical," a move that "confirms expertise," defines a problem to be addressed, and situates a boundary between those empowered to "diagnose deficiencies in others, and those who are subject to expert direction."[62] It is thus crucial to tease out the particular technopolitical configurations or networks, and their constitutive geopolitical and developmental moments, specific to each project and place that were created (and continue to be) and maintained in the service of large-dam proliferation and river basin development.

The technopolitical nature of large dams and the idea that dams and basin schemes are "in the service of" larger aims raise an important normative question, although it is a theoretical concern as well: Did the dams and river basin schemes that will be described in these pages, and the numerous others that emerged from Cold War technopolitical networks, accomplish their geopolitical goals? Did they in fact assist the United States in its effort to exert global hegemony and undermine Soviet inroads in the Third World? Moreover, did they accomplish their goals of fomenting economic development, industrialization, and enhanced agricultural production? At first glance, my case studies suggest a negative response. One of the projects (Yangtze Gorge), which was not initiated until four decades after it was first proposed, was originally put forth as part of an American policy toward China in the early 1940s that ended in catastrophe from a foreign policy perspective. Another two projects (on the Litani River in Lebanon and the Blue

Nile in Ethiopia) were eventually constructed, but quickly receded into history as relatively minor projects without any tangible political or economic benefits. A project that was to have remade the material and ideological landscape of Southeast Asia (the Pa Mong Dam on the Mekong River) was never built. But whatever the geopolitical and developmental intentions of these projects, they still mattered, and continue to matter, just perhaps not in the ways that their supporters imagined. This is a key point of James Ferguson's masterful analysis of development in Lesotho over the course of the later twentieth century: intentional plans, while important, are never important "in quite the ways the planners imagined." Rather, the "unintended outcomes" of development interventions are comprehensible within circuits of power and control that transcend those of the governments and experts that dominate the development apparatus.[63] The cases presented here offer a window into how developmental outcomes—in this instance, large dams and comprehensive river basin development—mutate in unforeseen ways under the influence of geopolitical and technological forces operating across several spatial scales. I argue that while many of the projects envisioned by the Bureau of Reclamation and their political overseers in the State Department "failed" in terms of their original technical rationales, they helped advance a technique for managing rivers that subsequently became hegemonic as a means for underdeveloped regions to develop their water resources.

Placing Dams in Space and Time

This book is constructed around a series of assumptions about how to translate the history of the geopolitics of large dams into something digestible and revealing. This translation involves a careful consideration of how research is designed, how information is collected, and how that information is relayed to readers. Here I lay out what I see as key dimensions of *how* this research has been carried out. These dimensions include my assumptions regarding the particular times and spaces that form the central focus of the book. I also feel compelled to offer a number of warnings—perhaps better described as "conceits"—regarding the book's scope and its epistemological overtures.

It is perhaps inevitable that a historical study grounded in geography is also a meditation on scale, and on how we think about both temporal and spatial scales when conducting research on complex nature-society relations such as those embodied by the global proliferation of large dams. First, it is worth noting the process behind selection of the book's temporal span, which runs approximately from the early 1930s to the middle 1970s. While forty years is an absurdly protracted period to assess the complexity and sheer magnitude of the mid-twentieth-century proliferation of large dams, it might arguably be said to be the "golden era" of dam building.[64] The year 1933 has a very specific connotation. It was at this time that the Bureau of Reclamation initiated construction of the Hoover (then called Boulder) Dam on the Colorado River. Hoover's construction "opened a new frontier in water resource development" and demonstrated to engineers, government officials, and the general public the efficacy of generating vast amounts of electricity by impounding a river. Perhaps less visibly, it likewise demonstrated that political backing for large dams was as vital as technological know-how in bringing these projects to fruition.[65] It was also in 1933 that the United States Congress enacted the Tennessee Valley Authority (TVA) Act. Quite rapidly, the TVA became a model for the world not only of what was in theory integrated river basin development, but also of river basin governance and regional economic development.[66] Hoover provided a lasting image of humanity's professed mastery over rivers and a potent producer of electricity to be used for economic growth, while the TVA presented an enduring example of how to put dams and other water infrastructure to work in the rational development of river basins. This study's ending date, roughly 1975, corresponds to the United States' withdrawal from Vietnam, from engagement in Mekong River basin development, and from international water development activities more generally. It by no means signaled the cessation of the Cold War, but it certainly marked a lessening in the US commitment to use technical assistance as a means of promoting American hegemony.

Returning for a moment to this book's genealogical starting point, Hoover Dam and the TVA can be cast in a different light if we think of dams and basin-oriented development programs as associations of actors rather than as stand-alone objects. Both were profoundly imbricated in a series of social, political, and economic processes that were quite specific to the United States of the 1930s. In other words, both were the products of a very specific set of historical conditions. The global economic crisis being experienced in the United States at the time was a critical factor in promoting the role of the federal government in massive public works schemes, a role that otherwise would never have materialized or would have been greatly scaled back.[67] Hoover was the first dam to illustrate that dam building on a massive scale was possible, in an engineering sense, and that it could provide an enormous supply of electricity. In several senses, the Hoover Dam is representative of the type of technological knowledge that revolutionized dam building in the twentieth century. I will revisit this topic in later chapters, but the technological innovations of Hoover—its design, the types of material (e.g., concrete) used in its construction, the techniques required to create a unitary impoundment—set the stage for a triumphalist period of global dam building. More broadly, Hoover established a pattern of activities and stabilized a number of relations—within society and between humans and nonhumans—that were equally important for the proliferation of large-dam technologies and expertise. Rivers became sites of production, in this instance production of electricity, that were perceived essentially as vast, untapped engines of economic growth and industrialization. Dams produced a commodity, electricity, that could be bought and sold and was critical to industrialization and economic growth. Rivers were never viewed in quite the same way. In a similar fashion, the TVA and the model of river basin planning and development it came to symbolize also endeavored to stabilize a certain set of relations, albeit with less "success" than the concrete structures that were so crucial to its social and economic ambitions. The creation of the TVA and the subsequent activities in river basin planning that it coordinated established an institutional and organizational setting that was the object of admiration and eventually imitation the world over, and one that was greatly abetted in its dissemination by official technical assistance packages such as the US-initiated Point Four program created in the aftermath of World War II.[68] As we shall see, the relations that these settings sought to generate involved political-economic and biophysical processes, but ones operating at quite different temporal scales than technological processes. Thus, while the events in this book range across several decades, seeing these events as a single historical strand belies the complexity of how the different processes that make up technopolitical networks originate and evolve over time.

Indeed, the networks that eventually assemble and are assembled by large dams spark a rumination on time as an active, important force within technology-society-nature relations. For example, the time required to conceive of a large project, offer up designs, undertake all necessary hydrological and geological studies, muster financial resources, mobilize political support, and ultimately, build the dam is often years, if not decades.[69] Meanwhile, the economic, geopolitical, and biophysical dynamics on which the creation of large dams depends follow their own temporal avenues, ones that twist according to their own historically specific influences and contingencies. Every project featured in this book—whether sited on the Yangtze, Litani, Blue Nile, or Mekong—had to confront this temporal dynamism in one way or another. And, in a fashion underscoring the power of large dams, every large dam erected in the twentieth century has a socioecological and ideological influence that far outlives the expertise and political aims that produced it.[70]

The spatial dynamics of large dams are similarly complex, perhaps none more so than the relations between dam and basin. One of my central assumptions is that the technological details, and hence the materiality, of large dams are contingent on the characteristics of specific landscapes. The dam site is a fundamental unit of analysis, or spatial scale, in this book, invoking as it does the *place* (stretch of river, river basin, nation-state, and so on) that serves as the fulcrum binding together assemblages of technical expertise, capital, labor, hydrological and ecological processes, political decisions, a host of biota, and the materiality of the dam itself. Different types of channels and river systems imply to the engineer different types of dams—contingent on history (what technical and scientific understandings of impounding rivers allow at a specific time) and geography (e.g., soil conditions, discharge rates, channel dynamics, etc.).[71] For the Bureau of Reclamation, confronting a wide variety of geographical circumstances from the arid and mountainous regions of the Litani River valley in Lebanon and the Blue Nile basin in Ethiopia to the humid subtropics of the Mekong River basin, deciding what types of dams to consider and eventually select for design work were critical dimensions of its activity. Yet such decisions were subject to the (geo)political desires of both American and host nation governments as well. And these dam site processes themselves transcend the merely local to include flows and networks at the scale of the basin and beyond. Indeed, the connections between large hydroelectric dams and river basin development, as mentioned above, are indelibly linked in the rhetoric of water resource planners and engineers. In theoretical terms, as large dams assembled various kinds of technical, political, biophysical, economic, and symbolic networks throughout the twentieth century, river basins gave those networks a tangible geographical scale.

Which brings us to another scale of analysis, one that is too often taken for granted within the social sciences, that of the nation-state. It is in this highly constructed space, ostensibly, that the decisions to identify a country's needs for water resource development, to engage technical expertise (often from elsewhere), to procure financing, and ultimately to build a project are debated and made. This scale and the local scale entwine in unanticipated ways, pointing out the relativity of spatial scale in the process. Although numerous dams conceived or constructed throughout the tricontinental world during the Cold War would not rank high among the world's titanic impoundments, the Kossou Dam in the Ivory Coast, the Mt. Coffee Dam in Liberia, and the Peligre Dam in Haiti (see the appendix) are all arguably the largest public works projects in their nations' histories. And finally, there is the utterly crucial scale of transnational flows—of expertise, materials, technologies, and in some cases, water. This is the geographical scale that is arguably the most important in explaining the proliferation of large-dam and river basin planning ideals during the Cold War, yet in many ways the most opaque. My aim throughout this book is to give each of these spatial scales—the dam site, the basin, the nation, the region, the transnational or "global"—its epistemological due while emphasizing their construction via technopolitical processes.[72] Moreover, in line with recent efforts in science and technology studies to transcend some of the spatial and temporal limitations of the network metaphor, my approach strives for an "ecological representation" of the heterogeneous entities that constitute technopolitical networks as they reside across different spatial scales and as they change over time.[73]

Finally, this book can be read as a series of conceits, in terms of its subject matter and the approach adopted by the author. Perhaps most obvious from an environmental and humanistic perspective was the absolute faith exhibited by dam builders and the promoters of large dams that flowing rivers could be effectively harnessed and mastered in the service of human needs. This conceit was expanded within the framework of river basin planning. Under this set of ideas, the notion of the single, multipurpose dam was vastly upgraded to include an entire series of dams, reservoirs, irrigation works, turbines, electricity transmission systems, and revamped agricultural systems that would be entrained in the service of economic growth, poverty reduction, and industrialization. This unshakeable belief in the desirability of water manipulation fed into deep assumptions held by the architects of US foreign policy throughout the 1950s and 1960s, who firmly believed that large dams and river basin planning, and technical assistance more broadly, would be perceived as an unequivocal good by recipient nation-states and thus assist American Cold War ambitions.

More reflexively, one of the irresolvable conceits at the center of this project is that it can adequately capture a slice of "global" or "world" history. There are significant problems with any such claim, not least the imperial character of a history that declares that its global scope is ultimately more comprehensive than histories of different geographical and temporal scales.[74] The alternative I have put forward here is that the history of the diffusion of large dams and river basin ideology is primarily a history of flows and networks, mindful of how these flows and networks linked to the particular scales and spaces at work in the geographical imaginaries of their architects. For this conceit of a history actually encompassing "the global" in its entirety—even if we bound the temporal scale—is mirrored by the conceit of development planners, foreign policy experts, politicians, and bureaucrats who imagined their efforts to promote technical assistance as truly global in scope. A work with different ambitions would transcend and subvert the global vision adopted here by paying more attention to the vagaries of the places on the receiving end of Bureau expertise.[75] And it would, of course, root out those largely unknown accounts of the peoples—whether Chinese, Lebanese, Ethiopian, or residents of the Mekong basin—who were living principally rural lives and experienced or otherwise encountered water resource development firsthand as they came into contact with Bureau field site teams.

It would also be a conceit to project too much in the way of intentionality or judgment on the Bureau experts whose subjectivities and actions are a major feature of this book. I was time and again impressed with the level of self-awareness and critical thinking displayed by the engineers and other specialists thrust into places and cultures for which they had little prior training or knowledge. To be fair, I was likewise dismayed by the often blatant prejudices and ignorance directed toward the government officials and people of the societies that Bureau engineers were ostensibly assisting in the process of water resource development. These prejudices reflect the broader racist and masculinist institutional structures deeply enmeshed within US foreign policy throughout the period covered in this book.[76] My own conceit is that I can convey a coherent interpretation of the geopolitics of large dams to readers without engaging more fully with questions of ethnicity, race, gender, and identity more broadly as revealed within the geopolitics of development described here. The masculinist character of the engineering profession—to my knowledge every engineer and water resource expert engaged in the Bureau's overseas activities was indeed a white male—has undoubtedly had a profound influence on how technical expertise was conceived and applied.[77] My hope is that others might draw resources from the histories presented here in order to investigate this important ground far more comprehensively.

A final conceit is that attention to historiography within my research design will be adequate to the research undertaking. Historical accounts of the global expansion of large dams and river basin planning do relate to material changes over time, no matter the difficulty in approximating the "actual story" of these changes through archival research; but these accounts are also constructs, dependent on the historical specificities and organizational contexts of source material.[78] I have tried throughout this book to avoid the urge to "go to history" as a tourist without a clear idea of how a particular chronicle might serve specific conceptual goals. Rather, historical research must maintain a constant dialogue between research goals and the efficacy of the archival materials that bring that history to light. The end result should be a plausible account of the historical processes under consideration, one that readers can assess and, if they are so inclined, reconstruct.[79] I also recognize that reliance on texts in the form of policy statements, memoranda, technical reports, and the like—both public and classified—is not without epistemological and methodological dilemmas. My methodological emphasis in interpreting and relying on these documents falls under what has been identified within critical geopolitics as the "agency concept of discourse," which—while a powerful conceptual tool for critically analyzing how geopolitical imaginations (or "narratives" or "visions") actively and problematically construct other spaces according to their own logic—is less attentive to quotidian practices and overemphasizes human agency.[80] Still, I have tried to transcend some of these limitations by presenting compelling accounts of both the actual practice of technical expertise in diverse locales and the geopolitical dialogues that facilitated such interventions.

Structure and Scope of the Chapters

The international activities of the Bureau of Reclamation provide a window into the profound linkages among geopolitics, technical expertise, and economic development on display throughout the twentieth century. Having provided an overview of the conceptual discussions that frame the book's subject matter, I proceed to an exploration of the genesis and scope of the Bureau's foreign engagements—along with the geopolitical and technical rationales for the program's operation—set alongside a series of detailed case studies from Asia, the Middle East, and Africa.[81] The cases presented in this book offer compelling evidence for the claim that technological interventions into the biophysical realm are never solely rational calculations, but are always intimately political.[82] As the following chapters (and the appendix) attest, there are dozens of relevant examples of countries, regions, and river basins where the Bureau of Reclamation played a critical role in disseminating the ideologies, practices, and knowledge associated with large dams and river basin development. The scope and length of Bureau engagement in a foreign setting varied tremendously, ranging from several weeks to over a decade (in rare instances).

The map represented in figure 1.1 is my effort to capture the wide range of places engaged by Bureau teams over the period 1933–1975 and to assess the relative levels of Bureau engagement in those places.[83] This image also serves to single out countries mentioned in the book. Needless to say, these locales are characterized by startlingly diverse historical-geographical conditions, and the similarity of the Bureau's approach in each place is a testament to the universalizing tendencies of technical assistance and water resource development as conceived within US foreign policy aims. An example from a team that worked in Nicaragua in 1951 (identified as "moderate" involvement in my illustration) offers a fairly typical window into the specific activities carried out in many of their overseas programs:

> The mission spent five weeks in Nicaragua in the early part of the annual rainy season, assembling and reviewing pertinent reports, maps, and unassembled information and making such field investigations of suggested projects as limitations of time and difficulties of travel would permit. A program for securing much needed basic data on streamflow and the topography of the area has been prepared, discussed with Nicaragua officials, and is recommended in the report.[84]

These kinds of reconnaissance reports were emblematic of many of the Bureau's early experiences in international development. Moreover, these reports were often requested by host governments with a rather specific purpose in mind, regardless of what the Bureau investigation might reveal. For example, Robert Newell, chief of the Nicaragua team, determined early in the mission that the "development of hydroelectric power appears to be the chief interest or at least the first interest of Nicaraguan officials."[85] Such preconceptions on the part of state officials in host nations of what the goals of water resource development should be did not, however, necessarily align with the Bureau's assessment of water resource needs and potentials. Nor did they reflect the geopolitical imperatives of the American state. Indeed, the global map represented in figure 1.1 can be read as a record of how technical assistance aligned with the United States' geopolitical engagements with Asia, the Middle East, Africa, and Latin America, but one that, when compared with the timing and duration of the Bureau's specific programs detailed in the chapters that follow, appears remarkably ad hoc and lacking in broad strategic vision. I ask the reader to keep these broad contours of the Bureau's international activities in mind when approaching specific chapters and arguments.

Chapter 2 focuses on the period that established the pattern of technical engagements between the Bureau and non-US states and societies that is reflected in later water development activities. Beginning in the 1930s, the Bureau received increasing requests from abroad for technical assistance, and a great deal of the support it provided became focused on the activities

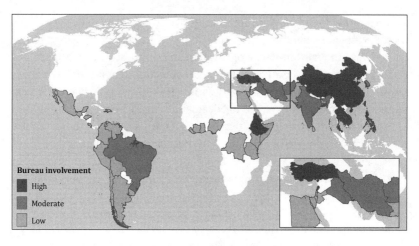

1.1. Map showing relative level of Bureau engagement in countries of the tricontinental world.

of a small group of notable personages. A key section of this chapter examines the use of technical assistance by the US government as a tool of empire building during the predevelopment era through an examination of the life and work of John L. ("Jack") Savage. Fresh from his pioneering efforts as chief design engineer on the Hoover Dam in the western United States, Savage served in his later career as an engineering consultant for numerous foreign governments, including those of India, Afghanistan, and China, during the 1930s and 1940s. Savage's consultancies and travels were approved and carefully monitored by US foreign policy officials, particularly in China during the critical years of 1943–1945. Using Savage's experiences as a template, the Bureau of Reclamation launched its Foreign Activities Office in 1950 as a response to President Harry S. Truman's call to aid the world's underdeveloped regions. Savage embodied what would later become a familiar icon of the post–World War II era: the "development expert" offering crucial technical advice to newly independent nation-states in Asia, Africa, and Latin America. Through his work, Savage and later engineers from the Bureau became materially and symbolically linked to the expansion of American hegemony in the middle part of the twentieth century.

Chapter 3 describes the period when the US State Department increasingly saw dams and river basin development as vehicles of technical assistance that, if used strategically, would demonstrate to current and would-be allies in the underdeveloped regions of Asia, Africa, the Middle East, and Latin America the superiority of American developmental and political approaches vis-à-vis the Soviets. This period also witnessed a formalization of the Bureau's role in international development. I examine the internal discussions within the American state about how best to organize technical assistance programs and, following creation of Truman's Point Four program in 1949 and its successor agencies, the Bureau's official entrée into the sphere of overseas development. An important dimension of the Bureau's institutionalization within the sphere of foreign policy encompassed debates over how economic and technical assistance might enhance the capacity of American business interests to increase their global influence and investment opportunities.

The central case study of this chapter is the Litani project in Lebanon (initiated in 1951), the Bureau's first intensive foray into technical assistance. The Bureau's experiences in the Litani River basin established an administrative and technological model of river basin planning that subsequent programs would follow, but they also reflect the numerous organizational and biophysical difficulties that harried nearly every major Bureau investigation in the tricontinental world. A key outcome of the Litani project and similar Bureau initiatives was the creation of what I call the "modern" river basin—a model that combined resource development through dam construction with more ambitious schemes of social engineering.

Along with the Litani River initiative, the Bureau's other major international effort of the 1950s was the Blue Nile investigation in Ethiopia. To provide institutional and political context, chapter 4 begins with a discussion of the rapidly growing scope of the Bureau's activities throughout the 1950s and 1960s and the role of technical assistance in expanding the liberal capitalist ideals of the American state. The chapter continues with a description of the Bureau's crucial role in Ethiopia's water resource development strategies, which stretched over a nearly two-decade period from 1951 to the late 1960s. The Blue Nile project helped to revive the Bureau's entire international program as it came under threat due to disinterest within the Bureau and lack of funding from Washington. Like the Litani experience, the Bureau's engagement with the Blue Nile basin exhibited a mingling of geopolitics, development aid, and technical assistance. The chapter highlights the Bureau's initial experiences with river basin planning in Ethiopia, the regional geopolitical considerations of concern to US officials, Ethiopian dissatisfactions with development of the Blue Nile, and finally, the actual outcome of the Bureau's investigations. The response of the Ethiopian government, and especially of Haile Selassie, to the Bureau's proposed development of the Blue Nile is particularly salient given the emperor's politically astute arguments for accelerated and more expansive American assistance.

Chapter 5 draws together threads presented in previous chapters (e.g., the technological and symbolic facets of large dams and river basin planning approaches, the tensions between technical expertise and geopolitical aims) and examines them using the case of the Mekong project, the Bureau's most intensive and longest engagement in international development. The Lower Mekong basin—shared by the mainland Southeast Asian states of Thailand, Laos, Cambodia and Vietnam—emerged as the focus of intense development interest to its riparian states and to international development institutions beginning in the 1950s. The period between the creation of the Mekong Committee in 1957 and the United States' disengagement from Mekong development planning in 1975 saw a potent blend of geopolitical imaginings, technological optimism, and faith in modernization theory that drove the proliferation of large dams and the idea of river basin development in mainland Southeast Asia. A key element in this story is the Pa Mong dam project, the focus of over a decade of study by Bureau engineers and experts and millions of dollars of US economic assistance, but which was never actually built. The Pa Mong project became the linchpin for development of the entire Mekong River basin and, in effect, helped generate an imagined geography of the Mekong region that resonates with more recent water development efforts.

The penultimate chapter (chap. 6) examines the contemporary geopolitics of large dams, asking to what extent the lessons of the Bureau's overseas endeavors can be applied to current debates over large dams, water development, and world politics. While the United States had largely vacated any role in the global promotion of large dams by the mid-1970s, the conception and eventual construction of dam projects continued throughout the tricontinental world with the assistance of Western liberal democracies, albeit without the more overt Cold War overtones influenced by American geopolitical designs. More recently, the Global South has been confronted with a plethora of reinvigorated plans for infrastructure development in major river basins, including the Mekong, the Blue Nile, and the Amazon basins. The global dam industry and the proponents of large-scale water infrastructure (including the World Bank) have championed hydropower development as a renewable and clean alternative to fossil fuels, although many scientists have reservations about their claims. While the Cold War no longer provides the dominant geopolitical frame through which dams are mobilized as ideological weapons, China's emerging role as global financier of large hydroelectric dams, particularly in Africa, demonstrates that the linkages among economic development, technical assistance, and geopolitics remain highly relevant to understanding world politics and the geographical transformations brought about through alteration of rivers. This chapter also proposes a "new" political ecology of large dams and river basin development that accounts for the changing geopolitical and environmental circumstances of the twenty-first century. A geopolitical analysis of dams enriches explanations of their continued salience to governments as developmental engines and of the emergence of a globally influential anti-dam social movement.

Chapter 7 brings the study to a close, offering a summary of the previous chapters' main arguments and a rumination on the possibility of developing a more deliberative and participatory approach to large dams as a development tool, while remaining cognizant of the multiple ways in which powerful geopolitical forces continue to encourage misguided efforts at water governance. Recent debates over large dams, as crystallized in the 2000 report of the World Commission on Dams (WCD) and its critics, have largely ignored the history presented here and have thus failed to account for the ways in which geopolitical forces can drive technological decisions while remaining largely opaque. Reimagining the goals of altering rivers requires, above all, engagement with and rethinking of the technopolitical networks that produce technological interventions and maintain their relevance over time.

TWO

Building a "World-Wide Fraternity":
The Bureau, China, and John Savage

Practically since the inception of the work of Reclamation by the Federal Government, progress in the construction and operation of the projects has been watched with ever-increasing interest by foreign governments, and every facility has been given their representatives for study and observation of the various problems connected with Reclamation.

—Arthur Powell Davis, Director of the Bureau of Reclamation, 1921 Fiscal Year report[1]

This chapter examines the emergence of technical assistance as a geopolitical tool within the apparatuses of the American state as it sought to extend its influence over the underdeveloped regions of the planet in the mid-twentieth century. This manifestation of technology and technical assistance as political devices is interpreted through the early work of the Bureau of Reclamation in "foreign activities" prior to its official recognition as an international technical assistance agency in 1950. As observed by Arthur Powell Davis, the United States' dream of reclaiming its arid western region through great irrigation works has long had a litany of admirers from beyond US borders. In many senses, the Bureau's technical knowledge was globalized before the formalization of its overseas work in the period following World War II.[2] From the time of Powell's statement until the announcement of Harry Truman's Point Four program of international development in 1949, the Bureau's experience with large dams and irrigation systems was admired and sought by a number of foreign governments. While this pre–Cold War geopolitical architecture differed in obvious ways from the era of bipolar world politics that was to come, the Bureau's activities were strongly if not exclusively guided by the United States' efforts to bolster its international influence, particularly in the critically important sphere of Asia. The network of geopolitical strategies and technical expertise that would define and shape the Bureau's Cold War activities was still nascent at this point. Yet certain key individuals presaged the Bureau's later activities and provided a blueprint, however embryonic, of much more intensive overseas engagements in the future.

The Bureau's overseas work in this period is exemplified by John L. ("Jack") Savage, who was stylized as the "first Billion Dollar American engineer" (due to the cost of the dam projects he designed and oversaw), while referring to himself as "just one of Uncle Sam's employees."[3] As this epithet suggests, Savage had a remarkable influence on the evolution of water resource development within the United States, having been the lead designer of dozens of major dams and associated power plants in the American West, including the Hoover (Boulder), Grand Coulee, Parker, and Shasta Dams and the All American Canal system. But it is Savage's less well-known exploits as an international consultant in places as diverse as Argentina, Palestine/Israel, Afghanistan, Australia, India, and (prominently) China during the 1930s and 1940s that provides the focus here. Savage's efforts provided the original model, or script, for the subsequent activities of the Bureau of Reclamation in promoting water resource development in the cases examined in later chapters. While we must be wary of ascribing too much influence to a single person, Savage's experiences as an engineering consultant—particularly his forays into China in the early 1940s—offer numerous insights into the origins of overseas technical assistance as a geopolitical instrument and the genesis of technopolitical networks that would eventually come to facilitate the proliferation of large dams. These insights also foreshadow later obstacles and challenges associated with the technopolitics of development and the Bureau's much more ambitious (and formalized) "foreign operations" during the Cold War.

Both the Bureau's and Savage's experiences during this early period raise a host of questions: To what extent was technical assistance and dam-induced environmental transformation a negotiation between the individual agency of engineers and the structural dynamics of American empire building? What do Savage's experiences in particular reveal about the messy and at times conflict-ridden relationship among technological expertise, practical geopolitics, and on-the-ground historical/geographical contingencies? This chapter proceeds with, first, a contextual overview of the Bureau's emerging influence as a prominent developmental agency within the United States and of its linkages to the imagined geographies of the regions targeted by the American state as zones of political attention. I then introduce John Savage as an agent of geopolitical power, using hegemony as a conceptual reference point (see chap. 1).[4] This focus on a single subject also raises important methodological challenges involved in historicizing the geopolitics of development. The rest of the chapter focuses on Savage's experiences overseas

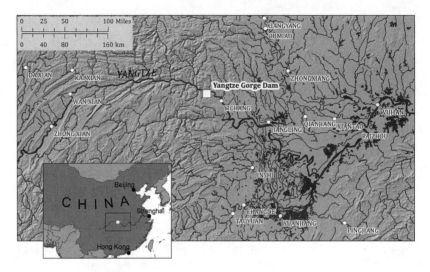

2.1. The Yangtze Gorge project location in its broader regional context.

working as a consultant for a series of non-US governments. His work in China in the early 1940s on the Yangtze Gorge project (fig. 2.1) is especially salient as a harbinger of the Bureau's future activities.

The Bureau Encounters the World

It is difficult to know precisely where to begin a historical overview of the Bureau of Reclamation's engagement with what came to be known within the agency as "foreign activities." Part of the inspiration for the creation of the Bureau was a report by the United States Geological Survey based on a visit by an engineer to India and Egypt in 1889. This report noted that many of the rivers of the western United States would in the future be used for irrigation purposes, and it pointed out the "similarity existing between these" and the rivers in the arid regions of South Asia. But what the Bureau encountered in the western United States, partly due to the extreme aridity of the landscape and its topography, led its engineers to conclude that developing effective irrigation projects would require "large hold-over reservoirs behind high dams, and the Bureau's efforts were devoted in this direction." This approach stood in contrast to that in India and Egypt, where low diversion dams were the norm for retaining and distributing irrigation water in the early twentieth century.[5]

Even in its early years, the Bureau was visibly part of a global network of hydro-engineering expertise that was circulated via scientific publications, professional conferences, and formal and informal communications across national boundaries. Before the late 1930s, however, the Bureau's international activities were quite limited, particularly given the irresolute nature of its domestic program during the first two decades of its existence. While its engineers succeeded in several major public works achievements that represented technical innovation of the highest order, the Bureau confronted financial hardships and reluctant support from the US Congress. Moreover, its primary aim of reclaiming arid lands for the benefit of small-scale farmers was compromised by land speculation, and a combination of "legislative requirements and political pressures" resulted in poorly conceived projects being rushed toward implementation.[6] Still, the Bureau's growing reputation in irrigation techniques led to short-term collaborations with counterparts in Australia and South Africa in the 1910s and 1920s as well as several advisory visits to Mexico in the middle of the 1920s to assist in the creation of a water resource agency for developing the country's arid lands through reclamation programs. As a result of this initial collaboration, Mexico inaugurated its National Irrigation Commission almost as "a mirror image" of the Bureau.[7] The 1920s and early 1930s also witnessed an ongoing series of key visitors and trainees at the Denver headquarters, among them individuals who would later take on prominent water resource development positions in the Philippines, Mexico, India, and South Africa. Throughout this period, the Bureau tracked the activities of governments and hydraulic entrepreneurs related to dam construction and irrigation development abroad. Folders related to "foreign activities" in its Denver archives contain numerous newspaper articles, requests for information, and similar communiqués from a broad range of foreign governments. For example, Bureau staff collected information regarding hydro-development in Africa, including Ethiopian plans from 1931 to build a dam on Lake Tana and the fantastical scheme of a German architect to create a "Great Lakes in Africa" by damming the Congo River and creating a "Second Nile" that would flow north to the Mediterranean Sea and turn the Sahara into a productive agricultural region.[8] This period of unofficial, informal interactions between the Bureau and its counterparts in other nations established a pattern of cooperation and mutual interest that laid the groundwork for the more formal relations that would characterize the coming decades. It did so in part by solidifying and extending the Bureau's reputation as perhaps the preeminent water resource development organization in the world, with a set of knowledge and skills surrounding "the construction of high dams" in particular.[9] John Savage, by virtue of his role in designing the most impressive of those "high dams"—Hoover—became an ideal vehicle for the Bureau's growing ambitions to demonstrate its technical prowess to both domestic and international audiences. Somewhat incidentally, these ambitions served the geopolitical interests of an American state intent on becoming a global power.[10]

"Just One of Uncle Sam's Employees"

John Savage served a minor but nevertheless noteworthy role as an agent of America's hegemonic ambitions throughout the expansion and subsequent consolidation of American geopolitical power in the middle of the twentieth century. Hegemony, put crudely, is an exertion of power by convincing others that what is beneficial to you is actually beneficial to all.[11] The idea of hegemony is useful for interpreting the Bureau's role in expanding American geopolitical power for several reasons. First, it directs attention to the fact that American hegemony is an extension of American society as much, if not more than, of the American state.[12] Perhaps more so than in any other nation-state, a broad range of American society—including political elites, tycoons, social reformers, and members of the general public—perceived technological prowess as key to, if not definitive of, national identity. The Bureau, in its domestic activities, was arguably at the vanguard of this world view, and John Savage "established and sustained" the Bureau's "stature as the world's most competent water agency."[13] As pioneer of the Bureau's general approach to overseas technical assistance, Savage spearheaded a model of water resource development that reflected a uniquely American faith in technology's power to improve human well-being.[14] Second, hegemony in the sphere of world politics has always been more or less an "outcome of assent and cooperation more than direct coercion,"[15] and this was particularly true of America's approach to the nascent "Third World" in the mid-twentieth century. The economic and technical assistance represented by large dams and river basin planning (and by the Bureau experts tasked with transmitting this assistance) embodies this "assent and cooperation" perspective and was explicitly designed to demonstrate the superiority of an American way of governance and life. This observation is borne out in the case of Savage in China presented here and in the instances of Bureau interventions in (for example) Lebanon, Ethiopia, Pakistan, the Philippines, Southeast Asia, and many other places in the 1950s and 1960s. Finally, efforts to expand American hegemonic power—during the Cold War and after—constituted a heady concoction involving the "frontier" character of the American economy (which involved expanding markets in "emerging" nations and societies), its cultural expression (the "ethos of the consumer-citizen"), and paternalistic dreams of modernizing the Third World under the rubric of development.[16] As noted previously, "Jack Dam" Savage and his work for the Bureau symbolized and presaged this amalgam of hegemonic elements.

While hegemony provides a useful analytical frame for examining Savage's place within broader networks of geopolitical power, it is less helpful in sorting through the methodological challenges of "peopling" the connections between geopolitics and technical expertise characteristic of America's hegemonic aspirations in the mid-twentieth century. Recent work in critical geopolitics and related fields highlights, for example, the crucial role of intellectuals of statecraft as key agents in the construction of popular geopolitics in the United States and elsewhere. Perhaps most importantly, work in feminist political geography has stressed the conceptual benefits of shifting the geopolitical lens away from the universal knowledge claims and "macro" approaches of much of geopolitical inquiry.[17] This shift implies moving our analyses beyond the ministries and departments of foreign affairs to examine the entire array of actors within the state itself that co-construct foreign policy—including "agencies dealing with economic and monetary policy, immigration and border security, citizenship and minority rights, foreign aid, and cultural exchange, among other spheres."[18] My focus on Savage and his work as a technical consultant for large dams in China and elsewhere—linked to and transformed within the geopolitical practices of the American state—is thus a logical extension of this broadening out of geopolitical perspectives on agency.

This chapter proceeds with, first, a brief account of Savage's life and early experiences with dam building in the United States and his notable technical achievements. I then turn to a detailed discussion of Savage's most renowned overseas assignment—consultations with the Chinese Nationalist government of Chiang Kai-shek in the early 1940s—in which I highlight the clear disjunction between technological and geopolitical desires within the American state. This clash of desires is represented most forcefully by the internal debates between staff members of the State Department and the Department of the Interior—prompted by Savage's conclusions regarding river basin development in China—over the efficacy of American involvement in development of the Yangtze Gorge (or Ichang) Dam.[19] Following the case study of Savage's Chinese experiences, I briefly examine the greatly scaled-up overseas activities of the Bureau of Reclamation during the 1950s and 1960s, when global geopolitical conditions relating to the Cold War has shifted sufficiently to create space for a much expanded view of what technical assistance could offer to the expansion of American power, which can be interpreted to some extent as a legacy of Savage's work.

The "Most Thrilling Work in the World"

John Lucian Savage's early decades seem rather uneventful; he grew up as part of a farming family in Wisconsin and completed his bachelor of science degree in civil engineering at the University of Wisconsin. By accounts he was hard-working, brilliant in his chosen field, and unassuming. He applied his engineering acumen to work with the US Geological Survey in Wisconsin during summer breaks from college. He joined the Bureau immediately after graduation in 1903 and was assigned to the Idaho division. For Savage, the question of the benefits, indeed the greatness, of dams and their transformation of rivers was a simple one. Here he describes his perspective regarding the Minidoka project on the Snake River, his first assignment for what was then the Reclamation Service:

> When I first went out to the Snake River Valley, I saw only a river, and a lot of wasteland. After the dam was up the land changed. It got water. Farmers moved in to work the soil. Crops grew. Then came villages and towns. That's why I think this the happiest, most thrilling work in the world.[20]

This linearity defined Savage's world view, in which, clearly, dams equaled progress. This attitude toward natural resources and water in particular also reflected the early American conservation movement and its association with Progressivism, which perceived the need to exploit resources efficiently while simultaneously maintaining a stock of those resources for future uses.[21] The late 1920s through the 1940s was a period of explosive growth of the Bureau as a bureaucracy—both in budgetary and staff terms—roughly corresponding to the height of Savage's career with the organization. Up until this period, the Bureau had struggled to find its mission in the face of competing demands from its ostensible clients, small-scale farmers seeking to reclaim the arid lands of the American West, and the array of business interests, urban boosters, and Washington politicians seeking to steer Bureau activities toward their own ends.[22] In addition, this was a period when Bureau leadership was both exceedingly stable and dominated by professional engineers who had great faith in industrialization, rational use of resources, and the scientific management of rivers.[23] John Savage epitomized this outlook. Although a public servant for most of his career, Savage worked for eight years (1908–1918) as a consulting engineer in Boise, after which he returned to the Bureau. Most of his time in Idaho, he recalled later, was "spent in the field on inspection and consultation problems,"[24] an inclination that characterized nearly all his engineering work.

From 1924 to 1945 Savage served as chief designing engineer for the Bureau and applied his considerable technical savvy to conceptualization of some of the most renowned projects in the agency's history of transforming the American West.[25] As noted above, he oversaw design and construction of a number of the major hydroelectric and irrigation projects. Savage's technical innovations are equally impressive. For example, he introduced the artificial cooling of mass concrete during construction of the Hoover Dam, which greatly reduced the time required for the concrete in the massive blocks to set and stabilize. He also pioneered the "trial-load method" for analyzing the appropriate mass and shape of arches to be used in dams by determining the inconsistencies between theoretical and actual stresses in gravity-arch dams, invented a critical type of needle valve that could withstand the tremendous hydraulic pressure applied to various apertures in dam structures, and contributed to the creation of numerical and graphical methods of vital use in dam design.[26] Savage's technical contributions to dam building were profound and provided him scads of legitimacy in his later encounters with foreign governments eager to undertake water resource development programs of their own.

Despite his deep commitment to the technical aspects of dam design and construction, Savage was apparently quite aware of the politicized character of dam construction in the United States.[27] As plans were proceeding for construction of the Hoover Dam in the 1920s, the commissioner of reclamation at the time, Elwood Mead, and Savage colluded to staff a "prestigious board of consulting engineers" with "friendly" experts who advised the Colorado River Board to support Savage's contention that the Bureau-approved dam design was preferable to previous iterations. The board eventually approved the Savage design for Hoover.[28] Despite this instance, Savage—according to co-workers and other observers—was an unpretentious, modest presence in his interactions with State Department officials, foreign dignitaries, and his fellow Bureau engineers. One reporter described him as looking "like any other 65-year-old bureaucrat earning $8,750 a year and content to plod along in the government groove."[29] Unable to rely on personal charisma or political acumen, Savage drew his power from a lifetime of knowledge regarding the construction of dams. An anonymous article titled "The Dams that Jack Builds," published in *Newsweek* on April 2, 1945, gushed:

> 1,000,000 years hence, if there are archeologists to dig out monuments of the remote twentieth century, they will find the nearest thing to pyramids of that day was erected not by a Pharaoh but by the obscure civil servant named Savage . . .

[who] was the greatest dam builder of an age in which hydroelectric power changed the face of the earth.

Savage apparently had an almost innate knack for identifying high-quality dam sites that often left co-workers in awe. According to one account by a fellow engineer, he had

> a natural bent for dams the way some people are natural athletes. I was on consultation jobs with him and saw him look over a river and its canyons, examine the pertinent data, then without hedging or "if-ing" recommend a precise site for the dam and specify the best type of dam for the job it was supposed to do. . . . He's the only engineer I've met who works on hunches. Several times I've heard him give decisions based on "my hunch." We believe his hunches. We've never known one to go sour.[30]

Savage's career with the Bureau highlights an individual who embodied the key traits of the idealized technical expert: virtuosity in terms of technical knowledge, a capacity to innovate, and an almost metaphysical understanding of how technology might best be applied to promote human welfare. He was a modest man, but one whose life mission revolved around the construction of large dams, and whose belief in the efficacy of these projects and their tremendous capacity to improve the human condition was unshakeable. Moreover, Savage personified a network of technical relations involving the conception of novel dam designs, the safe storage of vast amounts of water under tremendous hydraulic pressures, and innovative methods for speeding up the construction process. It was Savage's existence as a key node of technical relations—relations that also encompassed a breadth of experience from early work on the TVA and Hoover Dam to massive projects in the Columbia River basin—that made him so attractive to newly independent states keen on developing their own rivers along the pattern established in the United States.

"From a Weak to a Strong Nation": China and the Genealogy of a "Classic" Dam

During his later years as chief designing engineer, Savage began an active and impressively broad series of consultancies with foreign governments in the field of water resource development, a vocation he continued after his official retirement from the Bureau in 1945. In the words of a State Department press release, he was "at various times" a consulting engineer "on temporary detail from the Bureau of Reclamation in Puerto Rico, Panama Canal Zone, Honolulu, Australia and Mexico."[31] Just before his extended visit to China in the latter half of 1944, Savage conducted field surveys of several dam sites in India and advised the government of Afghanistan on water resource development.[32] In a foreshadowing of the later strategic and geopolitical orientation of the Bureau's technical assistance, Secretary of State Cordell Hull saw Savage's visit to Afghanistan as an opportunity to "convince the Afghans of the interests of this government in their problems."[33] Savage's actual consultations encompassed a range of activities, as disclosed in a 1943 communication from the commissioner of reclamation regarding Savage's work in India:

> In addition to working data such as tables, diagrams, compilations and manual information, the file of technical data will include: design data of all kinds; design examples relating to many of the important problems that will be encountered in the proposed work; design drawings that will serve as precedent for different alternative types of dam, power and irrigation structures; and reports showing methods of investigation and study in the economic development of river systems where irrigation, power, flood control, and navigation are involved.[34]

Washington clearly perceived the Bureau engineer's activities within a wider context. Before his departure for the one-year assignment in Afghanistan, India, and China, Savage received a memorandum from the State Department pointing out that "your work is expected to be useful to the war effort by aiding these governments in the control of water for food production and in the design of hydroelectric projects." In tasking Savage with this duty, the official notes that the State Department "has been influenced by your record of forty years in the field of civil and hydraulic engineering, including the past twenty years as Chief Designing Engineer" for the Bureau.[35]

John Savage entered China and the graces of the Nationalist government during a period of extended turmoil involving a global war, internal civil strife, and a highly volatile institutional setting. A brief historical sketch will help explain the (geo)political terrain that Savage entered in 1944 as an adviser to the Chinese government on water resource development.[36] Broadly, the US government throughout the 1930s and 1940s was intent on maintaining overt support for Chiang Kai-shek and his Nationalist Kuomintang party while preserving liaisons with the Communists at their base in Yenan Province. This period of China-US relations—and its morass of conflicting personalities and ill-advised policy decisions—is well covered in the work of a number of historians.[37] Savage was invited to China under the auspices of the Cultural Cooperation Program of the State Department, which had for several years initiated a series of activities designed to promote cooperation and mutual understanding on the part of Chinese and American societies, accomplished in part through an exchange program of experts in different technical fields. What awaited Savage in China—which he visited briefly in 1943 and again for a six-month period in 1944—was a regime ostensibly representing "Free China" that was corrupt and abusive in dealing with its own citizens, ineffective in military operations against the Japanese forces occupying Chinese territory, and made up mainly of a cadre of venal leaders eager for unequivocal US support in their ongoing internal struggles against Mao Tse-tung's Communist forces.[38] The Nationalist regime of Chiang Kai-shek (the "Generalissimo," as he styled himself and as he was referred to in State Department communiqués) was able to effectively enroll Savage's work on the Yangtze Gorge within a broader set of aims relating to the retention and potential expansion of ongoing US financial and military support. Savage's efforts in China, then, must be seen within this broader geopolitical context, a context that he seemed only vaguely aware of, but one that nevertheless shaped how his technical advice was received and acted on.[39]

We pick up the story in 1937, when some of President Franklin Roosevelt's staff articulated the notion that "the peace of the world is tied up with China's ability to win or prolong its resistance to Japanese aggression."[40] Shortly thereafter, Treasury Secretary Henry Morgenthau (the "most influential and active friend of China" in the Roosevelt administration) argued within foreign policy circles that Chiang's government could serve as America's "proxy" in Asia if provided sufficient economic aid. This position was confirmed by a US$25 million loan, which, although somewhat symbolic in character, consolidated US support for the Nationalists. Because of these and later actions, such as the creation of the Lend-Lease program in 1941—designed to allow the president to extend military assistance to any nation deemed to be vital to American interests—Chiang's Kuomintang regime perceived itself as an equal partner in a special relationship with the United States. Within this milieu, Chiang's brother-in-law T. V. Soong emerged as a vital conduit and confidant between Washington and the Kuomintang.[41] Soong was the director of the National Resources Commission (NRC), a powerful Kuomintang bureaucracy that coordinated resource development planning in China, and he became a leading proponent of development of the Yangtze. Perhaps more importantly, Soong had deep connections in Washington financial and political circles. Savage likewise bestowed tremendous technical and political integrity regarding the project

2.2. John L. Savage undertaking a field visit to the Yangtze Gorge, 1945. Courtesy of National Archives, Denver, Colorado.

within China; both Chinese engineers and political agents—such as Soong and Weng Wenhao—confirmed for the Nationalist government that Washington was serious about assisting in the development of the Yangtze and other water projects.[42]

By the time Savage arrived in 1943, the relationship between US advisers and the representatives of the Generalissimo was highly frayed. Still, Savage tackled his assignment regarding the Yangtze Gorge Dam with characteristic innovation and energy, and his activities included numerous field visits to the proposed site of the dam (fig. 2.2). This stretch of the Yangtze, situated near the city of Ichang in the province of Hubei, had long been seen by Chinese leaders as a resource of tremendous development potential. As early as 1919 Sun Yat-Sen had identified this part of the Yangtze as an ideal site for a dam that could provide both hydroelectricity and flood control. The upper

2.3. Artist's rendition of the Yangtze Gorge Dam as envisioned in John L. Savage's report, 1945. Courtesy of National Archives, Denver, Colorado.

reaches of the river are mountainous and characterized by deep canyons of a kind that for Savage must have recalled the Colorado River. Described as one observer as a place of "spectacular and sometime violent beauty," this region of the Yangtze has played a significant role in Chinese cosmology and literature, yet was also at the time surely one of China's most remote regions. The river's riparian zone provided moderately fertile soils for the production of wheat and other food staples, but the livelihood opportunities of the region's inhabitants were limited by environmental conditions, lack of nearly any social services, and rudimentary transportation.[43]

Savage's preliminary report on the Yangtze Gorge scheme is remarkable for the scope and scale of the project it envisioned. His concept of river basin development along this stretch of the river identified five potential sites for constructing the project's centerpiece, a massive "concrete straight gravity dam" roughly 225 meters in height (fig. 2.3). He saw the need for "twenty combined diversion and power tunnels," which could be located all on one side of the river or on both. The dam's power plants would produce an astounding 10,560 megawatts and store enough water to irrigate an equally impressive 10 million acres of "good agricultural lands." The huge reservoir of water behind the dam would provide "an excellent supply of clear domestic water for many cities and industries."[44] Additionally, incredible improvements to navigation would be realized should the project move forward. Savage envisioned a "single lock capable of lifting a ship 500 feet," which, given the enormous weight of the gates required to handle this lift, would require the digging of a shaft "through the mountain" to divert the river to the mountain's base, as well as machinery to "control intake and outlet of water from the shaft."[45] The total cost of the project was estimated at just under US$1 billion, and Savage projected net annual revenue after construction as US$154 million. In Savage's blunt words,

> The Yangtze Gorge Project is a "CLASSIC" [uppercase in original]. It will be of utmost importance to China. It will bring great industrial developments. . . . It will bring widespread employment. It will bring high standards of living. It will change China from a weak to a strong nation.

The reference to industrialization is an important one, since this issue eventually became a prominent part of the undoing of the entire project. It was at this point that Savage's vision for developing the Yangtze—and the Bureau's role in this development—encountered resistance.

The internal debate among Savage, staff of the Bureau of Reclamation, and State Department officials over continuing US involvement in the Yangtze Gorge initiative reveals many of the vagaries of technical assistance as a geopolitical tool that were to characterize later efforts of the Bureau's foreign operations program during the Cold War era. Most of these debates revolved around the Bureau's (and Savage's) desire to put its prodigious technical capabilities to work on a once-in-a-lifetime project and the State Department's more cautious approach to negotiating a diplomatic minefield and steering global geopolitical events. In September 1944 Savage cabled the commissioner of reclamation that the NRC desired American technical assistance for preparation of plans for the Yangtze Gorge project. Savage had discussed the matter with several key US officials in China, who expressed enthusiasm about the project. Savage's preliminary report—based on his research and fieldwork in China from May to October 1944—on development of the Yangtze was delivered to the US embassy in Chungking in November 1944.[46] In mid-November of 1944 the Department of the Interior suggested to Savage that the NRC would be advised to make the formal request via the American embassy, and that the request should clearly state that the Bureau of Reclamation should be designated by the president to provide assistance to the NRC. Following a flurry of communications among Savage, the Bureau and its parent Department of the Interior, American embassy personnel, and affiliates of the State Department, US officials acknowledged that the Chinese Nationalist government had officially requested that the Bureau be appointed to design a dam on the Yangtze and, "when funds are available, to construct the Yangtze-Gorge Project outlined in Mr. Savage's preliminary report."[47] Had the project gone ahead, it would have easily set records as the most expensive and massive engineering feat on the planet to date.

By March 1945 officials with the Foreign Economic Division of the State Department were questioning the legal mechanisms required to have the Bureau engage in a contractual agreement with a foreign government. More importantly, the United States began examining the project's economic (and political) feasibility in greater detail, and "China hands" in the State Department did not like what they saw. According to one legal expert, "even if the preparation of Mr. Savage's final report were to be undertaken by the Bureau, the action of the Department in connection therewith cannot be considered as implying approval of the project in relation to China's industrial needs and requirements, or any sort of commitment in regard to assistance in financing the project."[48] Despite such concerns, the Bureau, the State Department, and representatives of the NRC in the United States delivered a draft agreement for review to the secretary of state in early March 1945. After reviewing the document, the State Department stopped short of approval "on the ground that it would be economically disadvantageous to China since it would tie up capital in a development of power beyond the capacity of Chinese industry to utilize within the foreseeable future."[49] Moreover, it was clear to State officials that the Chinese government had not considered how this project would integrate with the "general progress of industrialization" and that China had little demand for the "vast quantity of power" the dam would generate.[50] These sentiments became the primary point of disagreement between the Bureau and State and the central stumbling block to initiating the Bureau's investigations of the Yangtze Gorge. Most importantly, a foreign policy official reiterated at a later meeting with Bureau staff that the State Department was "very much afraid that the preparation of the final report would give the Chinese the idea that the United States would finance the entire project" and felt that the "Chinese have been oversold" on the idea of the scheme.[51] While the planning of the Yangtze Gorge dam was by 1945 assembling an impressive array of technical knowledge (in the form of John Savage and his Bureau colleagues) and financial and political calculations (in the form of State Department officials), this technopolitical network was in no way preordained given the numerous institutional roadblocks it encountered.

Despite the misgivings of State, the Bureau remained fairly certain that its plans to produce a series of preliminary reports for the NRC regarding the development of the Yangtze would be approved. A good deal of that optimism rested on the confidence the Bureau had in Savage's legitimacy to carry forward the argument for Bureau involvement in the Yangtze project. For example, Secretary of the Interior Harold L. Ickes highlighted in a letter to State the numerous technical skills demonstrated by Savage during his stint in China under the State Department's Cultural Cooperation Program. Ickes extolled Savage's work and prestige, making it clear that all work on the reports would be "carried on, under the immediate supervision of Dr. Savage, by engineers regularly employed by the United States and Chinese engineers sent to this country by the Chinese government." Moreover, Ickes was adamant that the "Bureau will incur no financial obligations" as a result of participating in these reports, and that it would be in the Bureau's, and by extension, the US government's best interest to partner with the Chinese.[52] The Bureau, Ickes believed, would benefit in tangible ways from the Yangtze Gorge collaborations. Although the development of the Yangtze basin would present "many of the problems . . . similar to those which were encountered in the planning of Boulder [Hoover], Grand Coulee and Shasta Dams," the proposed dam and power development would "be much larger" and would thus afford Bureau engineers an opportunity to deal with "more intricate engineering problems" and apply this knowledge to the Bureau's postwar domestic program. Using a rationale for the Bureau's involvement in "foreign" activities that would become a hallmark of later engagements (see chaps. 3 and 4), Ickes contended that actual construction of the project would "furnish a market for the sale of great quantities of equipment and fixtures which American manufacturers will be glad to supply."[53]

Moreover, the response by the Department of the Interior and the Bureau to the State Department's recalcitrance to the Yangtze Gorge scheme in the face of perceived economic hurdles was swift and pointed, an indication of the significance they saw in the Bureau's participation. Michael Straus, the Bureau's assistant commissioner stationed in Washington, who as commissioner in the early 1950s would become a key architect of the Bureau's expanded foreign operations, contended that the State Department's "economic judgments" about the Yangtze Gorge project reflected the same mistaken arguments made against "domestic American hydro-electric developments." In Straus's experience, "industrial and agricultural development inevitably results from the availability of cheap hydro power and irrigation water and other benefits from multiple-purpose water control projects." Moreover, said Straus, "it is in the national interest of the United States to proceed with assisting China and preparing the technical reports essential prior to development of the Yangtze Gorge," since if this assistance were refused, "the task would then be done by others than United States citizens, probably not so well, inasmuch as nowhere in the world is there the wealth of experience enjoyed by the Department of the Interior officials in this field."[54]

And no one embodied this "wealth of experience" more than Jack Savage. Savage—in a rare revelation of his political thought—also reacted with frustration to the State Department's efforts to inhibit Bureau involvement in the Yangtze Gorge project. Upon hearing about State's unenthusiastic response, Savage wrote to Straus in late June 1945 that he found it "difficult to understand why our Government is appeasing Russia and at the same time kicking China around as usual." Moreover, Savage asserts that his fuller report on the Yangtze Gorge project—which was in transit from China to Washington at the time—would answer foreign policy concerns over the Yangtze project in "seven different languages." Savage goes on to praise the technical and developmental virtues of the project. He concludes on a sarcastic note:

> Still another phase of the project that apparently has escaped the notice of the State Department is that construction of the Yangtze Gorge Dam is warranted for irrigation, flood control and navigation benefits, and probably for any one of these benefits alone. It is unfortunate that the billion dollar and the 10,500,000 figures for the completed project appear to be the only parts of my report that have been read in the State Department.[55]

Who would stand in the way of such enormous benefits? What Savage and some of the other Washington-based Bureau advocates were likely ignorant of, however, was the geopolitical backdrop to what, for Savage, was a rather obvious technical decision in the name of a "classic" dam project. As the war was winding down, unification of Nationalist and Communist forces had become a top priority of US policy toward China. After Harry S. Truman's assumption of the presidency in April 1945, preparations were started to dispatch George C. Marshall to negotiate with the Kuomintang and Communist regimes "for the purpose of bringing influence to bear both on the Generalissimo and the Communist leaders towards concluding a successful negotiation for the termination of hostilities and the development of a broad unified Chinese government."[56] Little came of these efforts. As of December 1945 American advisers in China saw the political situation there as "by no means firm," with Chiang Kai-shek in particular trying to outmaneuver his political rivals on questions such as crafting a national constitution and convening a broadly representative national assembly. By this time, armed conflicts between the Nationalists and the Communists had broken out in the north of the country.[57] Given the delicacy and uncertainty of the geopolitical situation, it is hardly surprising that US foreign policy officials perceived the Yangtze Gorge project to be of questionable benefit. Moreover, US embassy staff in China had concluded that "enthusiasm for this project" was "largely confined to certain Americans." One would assume that those

Americans included Savage. The United States, it argued, would be wise to listen to the "numbers of important Chinese whose judgment we value" who argued that the project would not be "economically feasible for China at this time."[58]

Despite this hesitation, the Bureau's lobbying for technical assistance succeeded to a certain degree. The State Department approved an agreement crafted by the National Resources Commission and the Bureau of Reclamation (signed in October 1945) for Bureau staff (including Savage) to provide technical support for dam construction in China and, importantly, for the training of several dozen Chinese engineers at the Bureau headquarters and laboratory in Denver. Savage, after retiring in 1946 at the age of 65, immediately turned around and signed a contract as a Bureau consulting engineer to expressly continue planning work on the Yangtze project. Although "substantial progress" on dam design had been made by the summer of 1947, Savage and the Bureau evacuated China amid a worsening civil war later that year.[59] In addition, the Chinese engineers sent to Denver for in-depth training in large-scale waterworks were recalled by their government, and a minor conflict arose over whether or not the payments received by the United States for their training should be returned.[60] The only project to emerge out of Savage's detailed study was the Upper Tsing Yuan Tung Dam, located on the Upper Tsing River—one of the Yangtze's smaller tributaries—40 kilometers upstream of its confluence with the Yangtze. Construction of the dam was initiated in June 1946 by the NRC and was abandoned in May 1947 as the civil war intensified and the Chiang Kai-shek government became effectively bankrupt. The numerous American engineers hired by the NRC, including Savage and other crucial personnel, returned to the United States as the civil war intensified.[61]

An "Essential Ingredient" of the Cold War

A conventional history might see John Savage as an influential individual who, by virtue of selflessness and specialized skills, sought to improve the world through dam building. There is a certain element of veracity to this perspective, and I do not want to underestimate or misinterpret the motivations of this quite extraordinary person. However, I argue that Savage was equally a key component of a larger technopolitical network, albeit loosely congealed at this point, constructed around the Bureau's expanding influence within the State Department's geopolitical designs. Savage established the prototype for what was to become an extensive overseas program within the Bureau of Reclamation, one that became an important element in US foreign policy and expanding American influence in the postcolonial world in the aftermath of World War II. Additionally, despite the fact that the Yangtze Gorge project envisioned by Savage and the Bureau was never initiated, Savage's and the Bureau's work undoubtedly set the technical stage for construction of the world's biggest dam—the Three Gorges—a full half century later. And as we will see in chapter 6, the Bureau's early activities in China augured the emergence of the Chinese state not only as the world's preeminent producer of large dams, but as the world's leading exporter and proponent of large-scale water infrastructure in the current geopolitical era of global water governance. Savage's presence no doubt substantiated the zealous belief of Chinese engineers in all-inclusive river basin development—with the intent of squeezing the utility out of every drop of flowing river water for electricity generation, flood control, and irrigation—that has resulted in over 50 percent of the world's large dams falling within Chinese territory.[62]

What differed, however, between the aborted efforts in China and what eventually became the Bureau's Foreign Activities Office were the broader geopolitical conditions. The emergence of the Cold War, coupled to President Truman's pronouncement of the need for the United States to assist the world's "underdeveloped regions" in the renowned Point Four segment of his inaugural address, provided fruitful ideological ground for the use of technical assistance as a geopolitical tool. In response to Truman's Point Four proposal and the creation of the Economic Cooperation Administration in 1949, Commissioner Michael Straus delivered an agency-wide directive in 1950 enumerating the Bureau's "basic policy considerations" regarding international activities. These policies, Straus indicated, were a direct result of "the National Government's foreign policies for technological assistance to various foreign countries" and the already "numerous requests regarding overseas assignments for various categories of Bureau personnel [which were] . . . expected to increase many-fold."[63] Straus, who served as commissioner from 1945 to 1953, was a staunch advocate and key architect of the Bureau's entrée into overseas development consulting and project design, noting in a later polemic that "American water development technology and 'know how,'" the "American concept of comprehensive river basin development," and the "vast Western Reclamation developments such as Boulder [Hoover] Dam and Grand Coulee Dam" had "seized the world imagination."[64] There is a clear line of transmission between Straus's strong support for the Bureau's full engagement in China—including the eventual construction of the Yangtze Gorge project if some blend of Chinese, American, and international aid funding could be identified—and his later promotion of the Bureau as the world's most visionary water development agency.

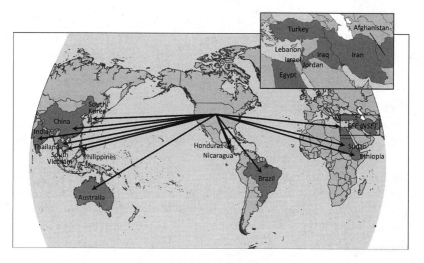

2.4. Re-creation of a Bureau map circa 1973 showing the multiple "foreign operations" of the agency from 1950 onward.

Indeed, the subsequent scope of the Bureau's foreign activities in the post–World War II era following its more piecemeal work in China is astounding, given that the history of these activities has registered so little in the eyes of scholars and the general public. While I return to this expansion in later chapters, it is worth noting that from 1951 to 1953 the number of people in the Bureau working on overseas assignments grew from forty to over a hundred employees serving in 21 different countries.[65] By the late 1960s, spokespeople could claim that the Bureau had "provided technical assistance in the field of multiple-purpose water resource development to over 108 nations in an effort to narrow the ever-widening gap in technology between the developed and developing nations."[66] The Bureau's network, vividly imagined in a 1973 map with Denver as the technical epicenter (recreated in fig. 2.4), could genuinely declare itself global. Additionally, there was wide-ranging awareness within the Bureau itself of its effectiveness as an apparatus of American foreign policy. Writing in 1953, George Pratt, director of the Bureau's Foreign Activities Office, first heralded the development of a "world-wide fraternity of those persons, institutions, and agencies, both governmental and nongovernmental, that are interested in the development of the world's water resources through methods of control, conservation, and use to the end that those resources may bring a higher standard of living" to the world's downtrodden. Pratt also addressed the unmistakable geopolitical stakes:

A hungry people may grasp at a communist straw and sell their free birth-right for a mass of potage produced in slavery. The development and use of the water resources of free countries is an essential ingredient of any program designed to relieve the world's hunger.[67]

Pratt and the Bureau's leadership thus perceived their efforts as not only a war against poverty and hunger, but by extension, a "crucial struggle" against communist forces. Underlying all the Bureau's foreign work was the basic geopolitical assumption that water resource development, and indeed. all technical assistance, was a necessary and vital component of the Cold War. The Bureau's staff were undoubtedly aware of the geopolitical facets of its mission, yet the organization simultaneously perceived large-dam construction and river basin development as worthy of advancement on technical and economic grounds. John Savage was aware of the geopolitical advantages that technical assistance in the form of water resource development would confer, but as his previous musings on "kicking China around" make clear, his perspective did not necessarily align with those of State Department officials. This cognizance of its geopolitical efficacy extended throughout the Bureau's work in later decades. While the Yangtze Gorge project foundered on foreign policy trepidations and civil strife within China, the Bureau's technical and ideological vision of large dams and river basin development as engines of economic development found more fruitful geopolitical terrain for its work in the 1950s and 1960s.

Conclusion

This chapter underscores one individual's role in promoting large dams and river basin development overseas and in inculcating the perception of these projects as paragons of objective technical design and construction. John Savage, by virtue of his impressive dam-building resume, assumed a position at the center of networks of technological expertise. In China, this expertise ran headlong into geopolitical networks that conspired against realization of constructing a "classic" dam in the Yangtze Gorge. By 1950 the world had changed significantly, and the "loss" of China was increasingly embedded within American discourses and policies of Cold War geopolitics. While it is unclear exactly how the Bureau's activities in China may or may not have influenced the geopolitical dynamics, there is no question that the American state perceived technical assistance as useful in terms of foreign policy objectives, particularly in those regions pigeonholed as "backward" or "under-developed." What remained for the United States, particularly the foreign policy officials of the State Department, was the challenge of finding a place for technical assistance in the broader effort to contain perceived communist expansion and bolster American hegemonic power. How could large dams and water resource development become—in the words of a 1949 treatise by Arthur Schlesinger, Jr.—a "weapon" in the geopolitics of the Cold War?[68] The disagreements and negotiations between Bureau and State Department representatives are notable for other reasons. The notion that the Chinese government—however inept and potentially corrupt—and members of Chinese society who might be directly and negatively affected by the Yangtze Gorge project should be consulted on how to best proceed with the dam was never entertained. This pattern, while shifting contingent on the specific geographical and political circumstances, was of course characteristic of nearly all large-dam and basin development projects of the period. This observation reinforces my contention that the State Department and the Bureau relied on imagined geographies, both in this case of the Yangtze Valley and of China more broadly, in their technopolitical calculations in the early 1940s. The Yangtze was reduced to a river of energy and industrial dreams that, if developed appropriately, would transform China into a "strong nation" within a startlingly short time frame. This vision of the Yangtze intersected with an understanding of China that was highly flawed, based as it was on certain expectations about internal political struggles. And underpinning both geographical assumptions was a deepening concern about the global dynamics at work at the end of World War II and a growing sense that communism was the primary threat to US hegemony in world affairs.

One of the goals of this chapter is to shed light on the role that technology has played in the global expansion of American power. Oddly, the "peripheralization of the technological dimensions of both nation-building within the United States and its subsequent emergence as a global power has been the norm rather than the exception in mainstream American historiography."[69] This is particularly remarkable given that American culture is so rife with technological artifacts, and in turn, that various technologies (e.g., steam engines, automobiles) have been so closely identified with American culture. One of the most notable features of large dams and associated water resource development processes is the fluidity of these technologies and their ability not only to travel geographically, but to transcend radically different ideological and cultural contexts.[70] It is clear from the discussions circling the potential investigations of the Yangtze Gorge that one of the guiding assumptions of the idea of developing the Yangtze basin was that such development would inevitably lead to greater levels of industrialization and, in fact, modernization. We thus see a style of thinking taking shape that directly foreshadows the much more systematic efforts—codified in part through Harry Truman's now famous Point Four speech—to "develop" the as yet unlabeled "Third World." As traced historically, the idea of *national* economic development—with the improvement of natural resources as a key component—had captured the imagination of nationalist elites in Asia, Africa, and the Middle East during the period of rapid decolonization coinciding roughly with World War II and its aftermath.[71]

The extension of American technical know-how in the area of large dams and river basin development expanded greatly in the era of Cold War geopolitics.[72] And the places where Bureau engagements were most intensive in terms of resources, engineers, and time—the Litani River in Lebanon (1952–1958), the Blue Nile in Ethiopia (1954–1962), and the Lower Mekong basin in mainland Southeast Asia (1956–1974)—corresponded to regions of substantial geopolitical interest on the part of US foreign policy officials. It is worth noting, however, that all of these engagements were characterized by high degrees of conflict (as in the China case) between the technical assessments of Bureau experts regarding what was feasible and the geopolitical objectives that State Department officials wanted these technologies to serve. In contrast to China, these later cases (as detailed in later chapters) often involved Cold War foreign policy experts seeking to rush through projects of questionable feasibility while their Bureau colleagues were more cautious in their approach. The technopolitics—the novel and complex blend of technological and (geo)political networks characteristic of the "development era" of the twentieth century—that ensued was not particularly effective in meeting either the geopolitical or developmental goals that drove the tremendous expansion of water resource development technologies and ideologies.[73] But the actual achievement of such goals is hardly the point. Irrespective of the intentions and interests of dam designers or the promoters of river basin approaches, the technopolitics being developed here ensured that these methods of transforming flowing water into something "useful" for economic development would proliferate regardless of the political and economic rationales and the socioecological consequences.[74]

And what are we to make of John Savage, who, according to one colleague, exhibited an "unfailing devotion to the task," represented an "extraordinary wealth of global experience," and brought a "gentle expertise and wisdom" to all his endeavors?[75] And what indeed was his role in "extending America" at the dawn of the Cold War? A response must be linked to America's cultural and political interpretation of technology as crucial to national identity and thus highly desirable as an export to the "downtrodden" of the earth. In reference to extending US interests and ideologies via water resource development, "Americans deemed their nation's historical development both unprecedented and unique, they also saw American institutions, ideas, and modes of organization as models for all societies." The experience of World War II, with the United States emerging as the sole global power, "heightened the moral and millenarian dimensions of this teleology" and strengthened exceptionalist claims whereby "technological prowess had come to be conflated with national virtue."[76] Savage, I argue, clearly embodied these values, and his experiences in China offered a well-defined blueprint for later efforts throughout the world's newly independent and "developing" regions. Savage was perhaps the ideal individual—moderate of temperament, a recognized expert in his field, seemingly impervious to personal ambition, and working toward some idealized "national" good—to bind together the national development aspirations of China's Nationalist government in the 1940s and the US State Department's desire to exert hegemonic, albeit subtle, influence on that government toward the attainment of specific policy ends. The Cold War's commencement in the years following World War II provided a nearly perfect window for the United States to use a quintessentially "American" technology—large dams and associated water infrastructure—and its ideological accompaniments—TVA-style river basin development that would ostensibly lead to poverty reduction and industrialization—to extend its influence throughout the postcolonial world. To the extent that Savage became the exemplar of the "American engineer abroad" and set the stage for hundreds of future Bureau engineers to engage in similar activities throughout the so-called developing world, he was certainly fully enmeshed in the geopolitical and technological networks of American hegemony and its expansion in the twentieth century.

THREE

"A Reclamation Program to Lead Them": The Bureau Goes Global

Gordon Clapp, commenting on the development potential of the Middle East as part of a United Nations mission initiated in 1949, espoused the notion of an "Arab TVA" for the rivers of the Middle East, in large part because it would "help put 652,000 unemployed Arab refugees to work on five public works projects to pioneer a long range Arab economic revival." This "Arab TVA" was to include the Jordan River, which drained an international basin shared by several Middle Eastern nation-states, as well as the Litani River, whose basin was the largest within Lebanese territory. Clapp, a former chairman of the TVA, called the Litani basin "one of the most magnificent for hydro-electric development in the world."[1] A scant two years later, following a global tour of the Middle East and Asia in 1951, Commissioner of Reclamation Michael Straus observed that the "underdeveloped regions" he had visited were "an area of new governments and an awakening people . . . [with] widespread poverty and concentrated wealth, with tremendous unharnessed natural resources." Unsurprisingly, Straus perceived reclamation—"bringing water to irrigate land and produce vital food and hydroelectric power to lighten the burden of toil and create industry"—as the "key" to their future prosperity. The commissioner surmised that leaders throughout the underdeveloped world were looking toward "a reclamation program to lead them from their troubles to a life of fuller promise and hope." The commissioner concluded, "What we call reclamation has gone global."[2]

Under what geopolitical and technological conditions did water resource development, particularly the development of river basins, "go global" and come to be perceived as a means to counter the "social ruthlessness" of the Soviet regime?[3] This chapter examines the deepening links between Cold War geopolitics and economic development to help explain the relatively rapid proliferation of the concept of river basin development and associated dam projects throughout the "developing areas" of Asia, the Middle East, Africa, and Latin America. Throughout the 1950s the Bureau's overseas engagements became formalized and institutionalized within the foreign policy apparatuses of the US government. The heart of the chapter is an examination of the Bureau's efforts in the Litani River basin in Lebanon, initiated in 1951. The Litani was one of the earliest and most concentrated of the dozens of Bureau involvements in international activities during the Cold War era (see the appendix). Its location within a region, the Middle East, of substantial political and economic interest to the US government helps explain why this otherwise nondescript project—which culminated in construction of the moderately sized Karoun Dam—became a focal point of US technical assistance and of regional geopolitical conflict. At a pragmatic level, development of the Litani largely failed to augment geopolitical alliances (for the US state) in any substantial way, and the hoped-for developmental benefits (for the Lebanese state) were problematic at best.

By invoking the concept of an "Arab TVA" and by explicitly linking it to the Litani River basin, Gordon Clapp also drew on an emergent understanding of the river *basin* as the geographical ideal for promoting water resource development. The Litani case thus prompts questions of how spatial scale is reimagined within the discourse of river basin planning and development. In effect, river basin development programs bound together narratives and practices representing global and regional geopolitics, national-level political and economic calculations, and a host of technical and environmental processes at basin and dam site scales. Put another way, the efforts to develop the Litani River along basin lines represent a particular kind of scale-making project, one that was promoted by the Bureau of Reclamation and was not particularly effective in the developmental or geopolitical terms intended by project planners—exhibiting a futility common to most "claims and commitments about scale." As Anna Tsing brilliantly clarifies, the "making of scale" within the realm of international development involves the conception and articulation of a "particular kind of view, whether up close or from a distance, microscopic or planetary." Different human agents engage in scale-making projects that imagine locality, nationality, or globality in ways that fit their social identities as well as their political motivations.[4] In the Litani case, as in all water resource development efforts and similar technopolitical interventions (see chap. 1), multiple actors were engaged in a variety of scale-making projects, ranging from the Bureau of Reclamation's emphasis on the basin, to the Lebanese state's attention to national and regional affairs, to the US State Department's commitment to a particular vision of global geopolitics. Alternative scalar visions—based on, for example, more localized understandings of basin livelihoods or the environmental conditions of the eventual project site—were simply ignored or neglected.

This chapter proceeds with an analysis of the geographical diffusion of the river basin development ideal as it evolved within the United States during the early twentieth century and of the impetus for its proliferation throughout the "underdeveloped" regions of the world in the 1950s and 1960s under the guidance of the Bureau of Reclamation. This analysis also demands attention to the shifting nature of the Bureau's role as a conduit of international technical assistance. I then consider the Bureau's activities in the Litani River basin in Lebanon in more detail, highlighting the entanglements among local socioecological relations, Cold War geopolitical networks, and the technical rationalities associated with the development of river basins.[5] Drawing from the Litani case, the conclusion considers how a greater sensitivity to the multiple scale-making projects at work in any endeavor involving technical assistance—but perhaps particularly in the case of large dams and basin development—might add to our understanding of the origins and maintenance of technopolitical networks.

The Bureau of Reclamation and the Evolution of the "Modern" River Basin

At first glance, it is hard to argue with the notion that the river basin—or its relatives, the watershed, catchment, and drainage area—is a natural physiographic entity. Indeed, the concept of the river basin "draws its strength from its 'naturalness' as a hydrologic and management unit," in spite of the fact that in many instances the river basin is also a social construct, in that it is put to use, especially by states and other resource management entities, to achieve certain political and ideological ends.[6] These goals, and thus how the basin is understood, can shift over time and space. Moreover, while manipulation of and control over water have historically been tightly connected to expressions of political and economic power,[7] the river basin as a unit of national development to be harnessed for the benefit of society is a relatively recent concept. The idea emerged among Western countries in the latter half of the nineteenth century when, spurred by utopian impulses and visions of controlling unruly rivers to maximize benefits to society, engineers, politicians, and scientists began to promote "transformation of the wild stream into the civilized river" along the lines of distinct drainage areas or basins.[8] Nascent river basin thinking and planning was evident in the

United States and several European nation-states by the late nineteenth and early twentieth century.[9]

None of these early efforts are comparable, however, to the scale and impacts of contemporary river basin development as imagined and applied over the course of the twentieth century. Interest in the basin as a unit of development began in earnest in the 1930s with the emergence of the TVA in the southeastern United States and its ambitious plans to stoke regional economic growth through flood control measures, navigation improvements, irrigation schemes, and hydropower development on the Tennessee River and its tributaries.[10] Such transformations required advances in technical knowledge and engineering skills as well as a particular understanding of human-environment relations that conceived of nature as a resource to be controlled and harnessed exclusively for the benefit of human societies.[11] The TVA subsequently became the preferred model for how nation-states— particularly those endowed with ample water resources—might best exploit their rivers to achieve economic and social goals.[12] Intrinsic to this reasoning was the understanding that river basin development was an integral part of modernization programs, in the United States and, particularly, abroad. "Development" required a diffusion of modern innovations—science, capitalist economic relations, technology, and entrepreneurship—that reflected institutional arrangements and societal values that were decidedly Western.[13] Although modernization theory was mentioned in chapter 1, it is worth reiterating that this theory—expressed succinctly as a "series of integrally related changes in economic organization, political structures, and systems of social values" modeled after the US experience[14]—served as a conceptual basis for programs of economic and technical assistance. Accordingly, American philanthropic organizations, government officials, and development planners "saw multipurpose development projects as a means to solve social problems abroad" and viewed river systems as "vast, untapped sources of potential energy that, if harnessed by technology," would provide a multitude of benefits.[15]

The TVA created the ideal of *modern* river basin development, wherein storing water and producing hydroelectricity—through novel water infrastructure technologies that allowed a previously unknown level of flow manipulation—would set in motion a set of highly integrated economic activities (e.g., agricultural production, resource extraction, industrial activities) to produce economic growth and higher employment levels for a specific region and its inhabitants, all coordinated via a highly centralized yet ultimately democratic authority. To become an effective agent of modernization, however, it was crucial that the concept of river basin development be able to travel, and this diffusion was greatly abetted by Cold War geopolitical conditions: the emerging power of the United States, superpower rivalry, and growing instability in the "underdeveloped" regions. However, backers of the TVA as a universal model of water resource development promoted its technological virtues while sidestepping, first, its close ties to US political and economic conflicts and, second, contemporaneous critiques of the program's shortcomings.[16] The TVA became a potent model for a generation of development professionals because admirers saw it as an approach "based entirely on a rational, apolitical, modular logic replicable elsewhere."[17] Yet from the outset, the organization was compelled to "tailor its mission to maintain its political legitimacy" and carry out its operations—including its system of dams—through careful "politico-technical manoeuvring."[18] Politically, the TVA was manifestly an outgrowth of Roosevelt's New Deal liberalism and commitment to large public works ostensibly for the benefit of underdeveloped regions such as the Tennessee Valley.[19] As such, conservative politicians and private utilities almost immediately decried its emphasis on government planning as a move toward state socialism.[20] In addition, despite claims by TVA officials that the program would promote "grassroots" democracy, a growing number of critics of the TVA pointed out that governance decisions in many areas of the valley were monopolized by TVA officials and local power brokers, that some of the program's initiatives—such as rural electrification—further marginalized women by neglecting domestic uses of power, and that the region's dams were starting to cause severe environmental problems.[21] Notwithstanding the critics' domestic concerns, TVA-style basin development became progressively more enrolled within modernization initiatives overseas and constituted an important part of foreign policy objectives for a succession of American regimes from the 1950s into the 1970s.[22]

While the TVA and its transformation of the Tennessee Valley became the global symbol of what river basin development might accomplish, it was the Bureau of Reclamation that spearheaded the global dissemination of the ideal of multipurpose development projects and basin-oriented planning through its international technical assistance programs. As noted previously, the Bureau of Reclamation, formed in 1902 and situated within the Department of the Interior, was primarily an agency charged with carrying out water development in the arid regions of the American West (see chap. 2). Before the completion of the Hoover Dam in 1935, the Bureau had focused on irrigating the arid and semiarid areas of the western United States, providing water resources to the region's growing agricultural regions. The building of Hoover, however, signaled a shift in the Bureau's role, positioning it to become a critical actor in transforming the West from a region of marginal economic activity into a major industrial and agricultural hub.[23] The technological and economic achievements of the dam did not go unrecognized outside the United States, as Bureau engineers became globally renowned (see chap. 2). With Hoover, the Bureau also perfected the idea of the multipurpose project: a dam that would meet several objectives at once, most commonly water storage for irrigation and flood control combined with the production of hydroelectricity. The logical next step, as far as the Bureau was concerned, would be to upscale the multipurpose project into basin-wide development involving multiple dams—thus gaining ever greater control over an entire river system's flows and providing the Bureau wider latitude to exercise its domestic water resource development goals. Moreover, hydroelectric dams offered a financial rationale for overall basin planning efforts because the revenues they produced offset the economically marginal bottom lines of Bureau-sponsored irrigation regimes.[24]

Ironically, the Bureau's plans for comprehensive river basin development in the Colorado, Columbia, and Missouri River basins in the decades following World War II foundered due to national and local political opposition[25] at the same time the model of the modern river basin was flourishing within its overseas programs. Harry Truman's aforementioned Point Four program and its focus on assistance to "underdeveloped" regions presented an ideal vehicle for international application of the Bureau's technical expertise. In his 1949 speech, Truman promised "to help the free peoples of the world, through their own efforts, to produce more food, more clothing, more materials for housing, and more mechanical power to lighten their burdens."[26] In later speeches, Truman envisioned "immense underdeveloped rivers and valleys all over the world that would make TVAs," if only their governments had available "somebody who knows the technical approach to their development."[27] As noted previously, Michael Straus, commissioner of reclamation at the time of Truman's announcement, proved to be an ideal institutional conduit for the model of the "water expert" personified by John Savage and his fellow engineers. What remained to be figured out, however, was the Bureau's precise responsibility as an international technical assistance agency and how it could simultaneously achieve geopolitical aims.

Empowering a "Global" Agency

Prior to the Bureau's official engagement with the Point Four program in Lebanon and its work on the Litani River scheme, its leadership engaged in a series of negotiations with the State Department regarding the precise character of the Bureau's foreign programs and their political, economic, and ideological objectives. These negotiations are instructive in highlighting the extent to which technical expertise on water resources—ostensibly objective—was enmeshed in the geopolitical trepidations of the American state. The negotiations (and tensions) between the Bureau and State Department officials covered a great deal of ground, but I highlight several aspects of them here: the scope of overseas technical assistance concerning dams and river basin development; the institutional arrangements necessary to facilitate Bureau service to foreign governments; and perhaps most importantly, the Bureau's stated need to maintain its reputation for technical proficiency in the face of the complexity of international engagements. I return to another critical arena of debate—the tension between the provision of technical assistance and the advancement of American business interests—in chapter 4.

The scope of the Bureau's "foreign activities," as first envisioned by its Washington-based leadership in the Department of the Interior, was astounding. In the late 1940s Secretary of the Interior Julius Krug and Assistant Secretary William Warne construed Truman's call for international development as an opportunity to radically scale up the Bureau's visibility, workload, and share of the budget. Following a briefing by State Department officials, Krug and Warne concluded that "participation in the [Point Four] program would call for a considerable increase of staff which could be effected [sic] over and above demands for the domestic program."[28] The list of countries included in this greatly expanded program of technical assistance reflects the geopolitical priorities of the time as well as the areas already identified by the Bureau as amenable to technical assistance due to previous contacts and requests, or as "ripe" for reclamation (see the appendix for additional details on these cases). In the "American Republics" (Latin America), major programs were envisaged for Mexico and Venezuela, and to some extent, for Brazil and Chile. Costa Rica had already requested assistance from the Bureau for comprehensive development of the Tempisque Valley. Similarly, Venezuela had requested the services of as many as 20 Bureau staff. In the Far East, Thailand and the Philippines were perceived as ripe for major programs. Thailand was expected to receive 30 US engineers because of its ongoing water development program (the Yanhee hydroelectric project) and because the Bureau had already trained dozens of Thai engineers (see the appendix). Other potential sites of intensive assistance (i.e., 20 or more Bureau staff members) in the Near East and Africa included Afghanistan, Ceylon (Sri Lanka), Egypt, Greece, Pakistan, and Turkey. India, on the assumption that "the production of sufficient food for the populace is probably its single greatest problem," was to receive 65 specialists to assist with the "dozens of great projects [that] are being designed and constructed." This plan corresponded to the rising conviction within India's political and scientific elite that dams and river basin development were critical to the country's modernization.[29]

Despite these grandiose ambitions (Krug and Warne anticipated that the Bureau would be able "to add sufficient personnel to its staff to have on the average 500 engineers abroad and some 500 trainees at one time"), some Bureau personnel realized the inherent limitations on the agency's ability to act as an international conveyor of dam-related knowledge. Giving his "frank opinion" of the expanded foreign activities proposed by Interior in response to the Point Four proposal, the Bureau's chief engineer noted that the recent requests for assistance overseas had nearly all been refused due to lack of funds. Some "Bureau men" who took foreign assignments had been "attracted by high salaries," but only a few had been willing to jeopardize their Bureau careers. The expansion envisioned under Point Four would almost certainly mean that the Bureau would "have difficulty supplying engineers for foreign missions without seriously retarding the domestic program."[30]

The assumed tension between greatly expanding the overseas program and sustaining the agency's domestic efforts never fully went away and was a theme in nearly all the Bureau's broader discussions of its international actions (as described in later chapters). Domestically, the immediate post–World War II era was a period of rapid growth of the agency's budget, personnel, and prestige within the broad contours of the American state,[31] and many staff members were presumably reticent about diverting its attention and resources to international endeavors. Early in the tenure of the Foreign Activities Office (which subsequently became a division), a number of Bureau personnel—including some of the top leadership—felt the agency would be better off without any obligations to conduct international technical assistance, which they felt was an unnecessary distraction from the Bureau's domestic program. As recalled by L. W. Damours, the division's most respected and longest-serving chief during the 1950s and 1960s and a highly accomplished engineer, the Foreign Activities Office had a number of "ups and downs" in its early stages. "There seemed," observed Damours, "to be a question within the Bureau as to its legitimacy and, at one period [roughly 1953] since it didn't appear to be willing to die, an edict was received from the Department [Interior] that it was to be strangled as quickly and quietly as possible." It survived in a rather "anemic" state, according to Damours, until the initiation of the Blue Nile investigations in 1957.[32] Most of the impetus for the Bureau's early activities in technical assistance abroad, and the resilience of its foreign mission in the face of internal criticism, unmistakably emerged from the corridors of the State Department and the architects of foreign policy.

Once this impressive vision of the Bureau's potential involvement in the field of international development had been laid out, the key challenges of the early years were generating institutional structures to fund these initiatives and attaining the appropriate bureaucratic approval from the US government. In the past, the chief engineer argued, the agency had acceded to requests for assistance out of a sense of "professional obligation, as well as a national obligation to extend a helping hand to those countries," and had been "forced to absorb most of the costs thereof as overhead, which in turn might be considered as a moral obligation stemming from the Bureau's pre-eminent position in engineering." In sum, he wrote, the Bureau is "now doing all that it can, and surely more than is legally authorized, in implementing the broad objectives" reflected in Point Four. The key, therefore, would be that all Bureau actions be paid for in advance and that funds "for administrative and preparatory work should be made available well in advance of actual participation in the technical phase of the program." Otherwise, he argued, the "lack of forethought and lack of funds sufficiently in advance to implement this program will result in great confusion and inefficiency" and reflect badly on the Bureau and the people involved.[33] Appropriate funding would also assist the agency in achieving the foreign policy goals it was being asked to take on, a circumstance not overlooked by Bureau leadership. A memo from Michael Straus reiterated the point that the Bureau's technical expertise served geopolitical objectives: if the Bureau received requests directly from a foreign government, that government would be advised "of the need for placing such requests through diplomatic channels."[34] Far from being the objective, impartial technical agency its staff often portrayed for public and political consumption, the Bureau was, at an early stage, fully cognizant of its position—as the "voice" of technical expertise—within the broader currents of American political and economic directions. Straus's directive was the Bureau's first effort to define its basic approach to foreign activities, and it offered a clear institutional path for applying the Bureau's expertise globally.[35]

In addition to institutional considerations, the Bureau's overseas forays raised questions about the distribution of the technical knowledge surrounding water resource development to "underdeveloped regions" and about who was most competent to wield this knowledge. Accounts from the Bureau's early years of technical assistance emphasize the organization's desire to defend and stabilize its expertise in water resource development for the power and legitimacy such knowledge conferred.[36] According to staff present at the inception of the foreign program in the early 1950s, it was critical that the Bureau's prodigious technical expertise in water resource development not be misused and delegitimated in the process of serving geopolitical ends. Bureau specialists lacked faith in the ability of State Department representatives and foreign officials alike to adequately assess a developing nation's needs in the area of water resource development. In the words of the Bureau's director of foreign operations, the "concept of an organization which engaged in the control, conservation, and use of water resources and the land resources in connection therewith is simply not one that is readily understood in the isolated cloisters of State Department diplomacy."[37] Two years earlier, the acting commissioner noted that "embassy aides, no matter how bright and industrious, are not qualified to interpret reclamation needs of countries in the terms that the Bureau requires to analyze requests and organize technical assistance missions."[38]

Bureau engineers also demanded that "qualified" Bureau experts undertake any assessment of water resource development potential as a prerequisite to more comprehensive involvement in a given locale. Indigenous assessments of water resource needs were perceived as untrustworthy. Following a trip to India in 1950, the acting commissioner "discovered what the country *really* needs from the Bureau" and that it was a far cry from "what it has requested and is now continuing urgently to request."[39] Most significantly, if the Bureau's technical superiority was not recognized and adequately prioritized in foreign programs of water resource development, there could be a loss of confidence in the Bureau's technological skills as well as unforeseen geopolitical consequences for the United States. In 1952 the Bureau's director of project planning noted that in the past several years "the Bureau has assumed the responsibility for the design of dams and other project works [overseas] without any knowledge of the hydrology involved." The director enumerated the ways in which a lack of basic hydrological data of reasonable quality, or in some cases, a blind acceptance of hydrological data provided by the country hosting the project, could lead to negative outcomes. For example, a "dam designed without an adequate spillway [a structure for channeling abnormally high flows over or through a dam] is a menace and may fail by overtopping," leading to "extensive damage and loss of life." Such outcomes "could not only hurt the Bureau's reputation but also severely damage the international good will that the technical assistance work is intended to build."[40] Ironically, Bureau staff members' professed outrage at the threat of their expertise being used and delegitimated for political ends in the context of foreign interventions overlooks the Bureau's own history as a dynamic political actor *within* the United States (mentioned above) and, in general, the highly politicized character of water resource development throughout American history. Over the course of the twentieth century, both individual dam projects and programs of river basin development were transformed in all phases of their implementation—from conception and design to construction and operation—through the efforts of congressional representatives, advocates of regional economic growth, agricultural interests, certainly Bureau officials themselves, and numerous other political actors.[41]

Despite concerns over its implementation and what it implied for the Bureau's international reputation in technical matters, Point Four breathed life into the Bureau's international activities. As part of establishing the institutional architecture of Point Four, an executive order issued on September 8, 1950, enlisted all federal agencies in activities related to foreign policy. As the order read, the United States, via Point Four programs, was "seeking to help other peoples help themselves by extending to them the benefits of our store of technical knowledge." Additionally, given that "Communist propaganda holds that free nations are incapable of providing a decent standard of living for the millions of people living in the under-developed areas of the earth," the Point Four and related programs were to "be one of our principal ways of demonstrating the complete falsity of that charge." Among the benefits to come from technical assistance listed in the order was the following: "Rivers can be harnessed to furnish water for farms and cities and electricity for factories and homes."[42] As an outgrowth of this order, the Technical Cooperation Administration was created in October 1950 as the primary agency responsible for implementing Point Four's goals. In anticipation of these events, the Bureau had created the Foreign Activities Office on March 30, 1950. As noted previously, prior to Point Four, the Bureau's international agenda was undeveloped and dependent on the interests of individual engineers known internationally as experts in water resource development (e.g., John Savage). By the early 1950s, however, this agenda had become more clearly defined and more closely aligned with the foreign policy agenda of the United States in the "underdeveloped" regions of the world. The Bureau's immediate response was to begin a series of technical missions that would establish its presence in the developing world for decades to come and would also provide the model for nearly all subsequent Bureau forays into the tricontinental world.

Finally, Point Four and its efforts to deploy technical assistance as a geopolitical instrument did much more than create new policies and administrative structures within the US government. Critically, it brought the discourse of development, and particularly "underdevelopment," into the so-called action agencies of the US government, those agencies that would later have tangible material and ideological influence on how development was carried out in the tricontinental world. Despite the numerous disagreements between the Bureau and its State Department contacts over funding, over the political implications of technological assistance, and so on, the two agencies shared a profound sense of the progressive character of their activities in terms of modernization and livelihood improvement. Point Four thus marked a critical point in the evolution of the American state as an imperial power, but one whose many appurtenances (prominently the Bureau of Reclamation) saw their actions not in an imperial, but in a developmental and modernizing light. This was certainly the case when the Bureau's technical assistance team arrived in Lebanon in 1951.

A "Pretty Good-Looking Project": Developing the Litani

In the Middle East and Central Asia throughout the 1950s and 1960s, the Bureau had more or less intensive engagements in Iraq, Iran, Jordan, the United Arab Emirates, Saudi Arabia, Pakistan, Afghanistan, Yemen, Palestine/Israel, Lebanon, and Turkey (see the appendix). Two officials of the International Bank for Reconstruction and Development (later known as the World Bank) argued in 1950 that the Middle East region's water scarcity made it an ideal locale for comprehensive river basin development.[43] Calling for a series of TVAs in the region, they noted that "great hopes have been attached to the control of the rivers of the Middle East as the means of increasing production and productivity, relieving population pressure, absorbing new population, and generally improving economic conditions."[44] As we shall see, great hopes were also attached to the geopolitical benefits such development was expected to catalyze. Here I provide a detailed account of how and why the Litani River basin, a relatively obscure basin amid many better-known Middle Eastern waterways (e.g., the Jordan, Tigris, Euphrates), became the focus of intense technological and geopolitical interest among an array of actors during the 1950s. I highlight how the technopolitical networks encompassing the Litani basin were animated by scale-making projects at global, regional, national, and less obviously, local and "project" levels. Ironically, the basin—the material scale at which modern river development activities were ostensibly directed—was reworked, ignored, and transformed by (geo)political dynamics associated with other scalar discourses.

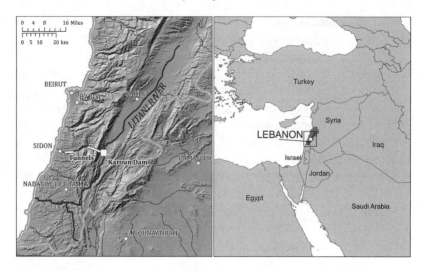

3.1. The Litani River basin, showing water projects discussed in this chapter.

The basin of the Litani River, which runs north-south along the eastern front of the Lebanon Mountains before turning west toward the Mediterranean Sea (fig. 3.1), became the site of one of the Bureau's earliest forays into foreign activities. The basin itself, covering an area of 2,168 square kilometers, can be divided physiographically into three sections: an upper basin set in the Biqa (Bekaa) Valley of eastern Lebanon situated between the Lebanon Mountains to the west and the Anti-Lebanon Mountains to the east; a middle basin beginning near the town of Qir'un (Karoun), a region of dry and rugged terrain; and a lower basin starting where the river's channel veers sharply westward before flowing through the Galilean uplands toward its mouth in the Mediterranean Sea. Precipitation across the basin is variable both spatially and temporally. Due to precipitation patterns and other geophysical factors, flows in the Litani are markedly seasonal—with high flows concentrated from January to April and low flows from July to October—and also vary from year to year.[45] As described in an early reconnaissance report on the region's development potential, the Litani River basin "is a rural region of rugged mountains and fruitful green plants. Its people make their homes in villages and small towns along the lower slopes of the mountains."[46] Some sense of this landscape is conveyed by a photograph from a 1955 Bureau report (fig. 3.2). According to Bureau team members, the basin's inhabitants practiced "primitive" agricultural methods and were thus (as was the case with so many developmental interventions during this period) perceived to be in dire need of modernizing technologies and

3.2. Litani River basin landscape downstream from Bureau-identified dam site. Source: US Bureau of Reclamation, *Development Plan for the Litani River Basin* (Beirut: Litani River Investigation Staff, 1954), frontispiece.

institutions. The basin's only major city, Zahle, had a population of 25,000. Even after construction of the water infrastructure described below, observers saw the basin as "on the fringe of Lebanon socially as well as physically."[47] From the onset of the Bureau's activities in Lebanon, the goal was to transform and modernize this largely rural, underdeveloped region through water resource development.

The idea of developing the Litani River basin was first broached in studies conducted by the French colonial regime in the 1930s. George Maasry, a prominent Beirut businessman, spent several years devising river basin development plans for the Litani, and all involved hydroelectric dams with a diversion component for irrigation. A 1948 report by Lebanon's Ministry of Public Works looked at the feasibility of some of the same infrastructure projects as the subsequent Bureau development plan (released in 1954), and the river's development potential was confirmed during the aforementioned UN survey mission in 1949.[48] By 1950 Commissioner Michael Straus had worked out the details of the Bureau's operations in Lebanon. At the same time, as noted earlier, he was actively advocating for a prominent Bureau role within the Point Four program and other US technical assistance activities. Straus's firm belief was that the Litani project "if initiated rapidly can serve in the Levant as a demonstration project for the Point Four program." He entreated Interior leadership to "vigorously present the Bureau's ability to carry through the project" to decision makers in the State Department.[49] The existence of previous studies on Litani development—the Bureau recognized in its 1951 reconnaissance report that its study "was largely in the nature of a review" of previous investigations[50]—combined with the strong support of Bureau leaders for overseas programs made the Litani an attractive place for technical assistance.

The work of the reconnaissance team, headed by Bureau engineer Robert Herdman, was initiated in April 1951. The writings of Herdman reveal several facets of the political and cultural aspects of Lebanon that foreshadow many of the challenges that Bureau experts would encounter in future overseas assignments. For example, Herdman observes the "arduous" nature of official protocol that required them to "be introduced to certain officials in the villages of the [Litani] valley." Moreover, the "usual American 'dash' and aggressiveness" needed to be "held in check" in the face of a "temperament and way of life greatly different from that of the United States."[51] From the onset of the mission to Lebanon—which included, in addition to Herdman, a ten-person team consisting of specialists in irrigation, agriculture, electrical engineering, geology, and hydraulic and sanitary engineering—the program was intended to be a "fore-runner of similar developments that might benefit other sections of the world."[52] The team was idealistic: the irrigation specialist emphasized to Lebanese officials that similar reconnaissance reports tend to "forget the people" by placing too much emphasis on construction.[53] Yet conflict arose over the seemingly incorrect perception that the Litani held an abundance of water and, as Herdman feared, between Lebanese bureaucrats favoring hydroelectric development and those promoting irrigation. Still, Herdman concluded that the Bureau team would be able to "come up with a pretty good-looking project" that would include "some irrigation and some village water supplies and a fair block of power."[54] What eventually coalesced was something rather different.

After the release of the reconnaissance report, the State Department agreed to fund a two-year study by the Bureau (1952–1954) into the development potential of the Litani River, and Bureau staff remained in Lebanon for several years after completion of the basin plan to serve in advisory roles. Before and throughout the period of the Bureau's engagement, the foreign policy apparatus of the United States targeted the Litani basin, and Lebanon more broadly, as important due to their potential geopolitical significance in the global Cold War and, relatedly, their links to regional hydropolitics. US foreign policy makers perceived Lebanon as a key component of a broader geopolitical vision focused on the strategic importance of the Middle East, one that in the early 1950s revolved around support for the nascent Israeli state while also maintaining an influential position among the region's Arab states, albeit in a more or less indirect fashion.[55] Under this general policy, Lebanon was singled out as a potentially effectual ally in the region. After his assumption of the presidency in 1952, Camille Chamoun was described in a State Department report as an "honest and industrious" leader who had visited various Arab capitals and "secretly tried use influence in promoting regional defense with West." Middle East experts in the State Department advocated the use of economic and even military aid to encourage political stability under Chamoun, given that "democratic government has a broader base in Lebanon than elsewhere in the region."[56] Technical assistance programs were perceived to be an important component of the overall geopolitical orientation toward the Middle East, and the Litani project was identified as an exemplar in this regard. In the same report cited above, State Department staffers highlighted the fact that the technical assistance program in Lebanon had "finally reached firm ground" and that it was the "largest staffed program in Arab [sic] states."[57] From an early stage, the Litani River, Lebanon, and the Middle East were components of a global geopolitical vision put forward by the US government as part of a Cold War strategy to contain the perceived Soviet threat.

The Litani's embroilment in regional hydropolitics revolved around Zionist and later Israeli state plans to enhance water security via diversion of the river's flows and international efforts to effectively settle the over 650,000 Palestinian refugees displaced by the creation of Israel in 1948. Studies from the 1930s and 1940s—the most prominent by Walter Clay Lowdermilk—asserted that the Litani's flows were underutilized in Lebanese territory (one study claimed only 14 percent) and were used in a succession of Zionist initiatives to argue for a tunnel or similar inter-basin transfer scheme to bring Litani waters to what was then the Palestinian Territory.[58] As the Bureau survey of the basin unfolded in the early 1950s, a series of confidential discussions within the State Department made it clear that the newly independent state of Israel also perceived access to the Litani as part of a broader Middle East "water for peace" program involving the sharing of the Jordan River.[59] For example, one American regional specialist noted that the "Department has received an Israeli note . . . calling for a regional application of Lebanon's Litani."[60] In the following year, the Israelis communicated to US officials their proposal for a water development scheme based on the "combining of Jordan and Litani waters."[61] Cooperative development of the Jordan River basin—coined the Johnston Plan after the special envoy designated by President Dwight D. Eisenhower to generate a comprehensive plan for water resource development in the region—had been promoted by the United States and the United Nations as a way to simultaneously stimulate economic development and encourage cooperation between Israel and the "moderate" Arab states in the region. For the Israeli government, securing long-term access to the waters of the Jordan basin and, if possible, the Litani was an exercise in sovereign power linked directly to its nascent project of nation building.[62] As mentioned earlier, US officials also hoped that development of the Jordan and Litani and the creation of additional "Arab TVAs" might ameliorate the "most urgent issue" of the "resettlement [of Palestine Arab Refugees] in Arab lands."[63] Thus the Litani was clearly implicated in global and regional geopolitical visions and policies, and this, as we shall see, to a significant extent drove the material transformation of the basin through the application of the Bureau's technical expertise.

From the outset of the Bureau's Litani program, which began formally in 1952, both Lebanese and American actors were aware of the politicized nature of technical assistance, and they tried to shape the Litani project to meet their specific goals. The US government sought to link economic and technical assistance to its global and regional geopolitical ambitions, albeit within a narrowly defined role for how development assistance might be used. As general policy, the Eisenhower administration reined in the scope of Point Four and other aid programs, believing that private investment was to be promoted over public capital supplied through grants and loans, and that "only governments that Washington perceived to be under direct threat from communism or deemed to be allies in the effort to contain it" should be qualified for direct US financial assistance.[64] In Lebanon, Point Four officials were reticent about delivering high levels of aid to the government; instead, US policy was "to discourage the Lebanese from looking for a handout, while encouraging them to find means of self-financing [development projects] and to utilize international banking institutions."[65] US embassy officials in Beirut were even more blunt about the country's geopolitical utility: "Since Lebanon cannot settle additional [Palestinian] refugees and has a limited amount of land which can be brought under irrigation, there is not as large an argument for economic aid as in some of the surrounding countries."[66] In spite of the State Department's modest opinion of Lebanon's strategic importance, American officials trumpeted the Litani intervention as a blueprint for technical assistance programs elsewhere, seeing the Bureau's planning activities as a critical means of disseminating American ideas regarding modernization and the value of geopolitical alliance with the United States. In 1953 the head of Point Four in Lebanon relayed to the State Department that the Litani River survey was the largest Point Four project in the country and that the United States had 48 technicians and 15 program direction staff working in Lebanon.[67] Even at this stage, the Litani development program resided within a contradictory space that encompassed US efforts to determine the political effectiveness of technical assistance in promoting its strategic objectives within the global Cold War context as well as an evolving regional geopolitical landscape where water resource development was a crucial consideration.

The Litani also figured prominently in the national development agenda of the Lebanese state, which was oriented around increased agricultural production, domestic electrification, and encouragement of fledgling industrial activities. Accordingly, the Lebanese state viewed the Litani project as an important benchmark for the veracity of US rhetoric about the promotion of economic development through technical assistance in the Middle East. For example, President Chamoun expressed to US officials in the early 1950s his concern over the lack of progress on the various development initiatives put forward under the Point Four program. The American chargé d'affaires in Beirut attempted to mollify the president by assuring him that the "object [of the Litani studies] was to prepare document which IBRD [International Bank for Reconstruction and Development] would accept [and provide funding for] without further technical study." Chamoun responded that he still wondered when the United States "would give Lebanon the kind of economic aid we [the United States] have freely given to even less friendly European countries and Turkey."[68] These kinds of tensions revolving around the pace of tangible water development projects lingered as the Bureau's activities progressed in Lebanon and in other countries as well, particularly in the case of Ethiopia (as described in the next chapter). For example, the State Department acknowledged in 1955 that the "Lebanese would like to receive economic aid on a much larger scale than at present" and would "like us to finance a significant portion of the Litani River development project."[69] The Litani, then, became a focal point for attracting aid that the Lebanese state perceived as vital for the national good, and government officials recurrently gauged the willingness of the US foreign policy apparatus to expand development assistance in support of its geopolitical aims.

Israeli designs on regional water resources continued to be a major concern of the Lebanese government. At an early stage of the Litani project, US embassy and Bureau officials in Lebanon recognized that any effort to locate a dam and irrigation infrastructure near the southern portion of the basin was "loaded with political dynamite because of any potential transfer of

Litani water into Israel water shed."[70] As Bureau experts explored different basin development options, a diversion scheme involving construction of a massive tunnel near the Karoun dam site to redirect the river's flows, referred to as the Bisri Tunnel (described further below), became a preferred option over more technologically and economically desirable components of the overall basin program. The Bisri diversion scheme had "many political and other advantages over the other schemes," enough, apparently, for the Bureau to advocate for its construction despite its awareness of many potential problems regarding the geology of the site and the overall economics of the project. The "major political factor" in moving forward with the Bisri scheme, concluded US officials, was "the conviction of Lebanese officials that all major developments should be located at a distance from the Israeli border."[71] The concerns over Israel appropriating the Litani's flows continued for the duration of US involvement in basin planning, and the Lebanese state's interpretation of regional hydropolitics actively shaped what eventually emerged as the preferred basin plan, and the resulting transformation of the Litani basin, in a quite specific way.

As the foregoing discussion emphasizes, Bureau engineers contended with myriad geopolitical undercurrents as they initiated their technical studies, executed a comprehensive river basin development program, and offered expert advice to their Lebanese collaborators throughout the 1950s. In contrast to the global-, regional-, and national-level concerns of the US State Department and its Lebanese associates, the development plan that finally emerged in 1954, at a total cost of US$830,000, squarely identified the *basin* as the most important scale for Litani development. Project engineer Robert Herdman noted that from its earliest study, the Bureau's goal was to "investigate the feasibility of basin-wide development," with the intent to "make maximum use of water for irrigation purposes" along with the construction of hydroelectric plants to supply electricity for domestic consumption and industrialization.[72] The Bureau's investigations also included a power market survey, identification of potential dam sites, a survey of water supply needs, and estimates of malaria and pollution control measures.[73] In line with the concept of the "modern" river basin that emerged following creation of the TVA and the Bureau's efforts in the American West, the plan assumed that harnessing the Litani through infrastructure and careful planning would inevitably contribute to economic benefits. Yet river basin planning and development also demanded a competent water management organization. The Lebanese government, on the advice of Herdman and other Bureau staff, thus established the Litani River Authority (LRA) in 1954. The Bureau estimated that the infrastructure components of the plan, designed to produce 170 megawatts of electricity and irrigate 20,000 hectares, would cost just under US$100 million and would take twenty-five years to complete.[74] The first phase of the project—which remains the only portion actually completed—included the Karoun Dam and reservoir, two diversion tunnels and power plants, and an irrigation project (see inset of fig. 3.1), at a total cost of US$35 million over six years.[75] Almost immediately, the Lebanese government signaled its intent to seek funding from the International Bank for Reconstruction and Development (the World Bank), which US officials had virtually guaranteed would be forthcoming.

What ultimately emerged from the Bureau's plan was a far cry from the basin-wide development conceived by project planners, and the reasons for this are directly related to scale-making projects at both regional and national levels. As noted above, the first phase of the Litani project involved construction of a series of dams, diversion weirs, tunnels, and a transmission system in order to accomplish the twin goals of hydroelectricity production (a total of 170 megawatts) and irrigation development (a revised total coverage of 18,600 hectares). Despite the expressed interest of both Bureau experts and the Lebanese government in "comprehensive basin development," a single large dam and associated infrastructure quickly became the focal point of the program.[76] This structure, the Karoun Dam, eventually created a reservoir of 180 million cubic meters of water, which was then diverted downstream through a five-kilometer pressure tunnel to a 25.8-megawatt hydroelectric facility at Markabi. The flows were then redirected by a low dam on the Litani through the 15-kilometer Bisri Tunnel to an additional 60-megawatt hydroelectric plant at Jezzine on the Awali River on the other side of the Lebanon Mountains. Initially, the water in the Karoun Dam's reservoir was scheduled to irrigate 5,700 hectares in the immediate area, with the hope that flows coming through the Bisri Tunnel would irrigate the semiarid coastal plain and provide water to Beirut residents.[77] Eventually the World Bank awarded a loan of US$27 million to finance the project, contingent on some support from the Lebanese government.[78] Construction began in 1957 and was slated for completion in 1960, although for the reasons described below, completion of the Karoun Dam was delayed until late 1965.[79]

The experience with the Bisri Tunnel underscores the active role taken by environmental processes in altering technological and political trajectories of economic development and the unexpected contingencies that characterize nearly all development interventions. The tunnel was in part an outcome of the Lebanese government's desire to divert water away from its politically sensitive southern region, which in theory would lessen the basin's attractiveness to Israeli planners. As noted above, the tunnel was designed to travel underneath the Lebanon Mountains and, ultimately, to provide sufficient flows for delivery of water to Beirut—and for a hydroelectric power station—on the other side of the range, as well as some irrigation development. By August 1960 the construction of the 15-kilometer Bisri Tunnel had become "a nightmare for everyone involved."[80] The problems were extensive: poor rock quality required extra boring, bracing, and lining of the tunnel; several cave-ins halted all construction; and water poured into the tunnel, occasionally at rates of 95,000 gallons per minute and pressures of 850 pounds per square inch. The worst came when massive quantities of sand leaked into the tunnel, virtually filling an entire 3-kilometer stretch. In the words of an industry magazine at the time, "All efforts to remove the sand have been unsuccessful." By this time the Bureau's experts had abrogated any responsibility for consulting on the project, yet the "problems of geology and geohydrology" became so complex "that a board of internationally renowned engineers and contractors [including former Bureau staff] was called in to study conditions and suggest remedial action."[81] These technical delays and complications were compounded by subsequent political maneuvers within Lebanon in the 1960s that further altered the original vision of the Bureau. The project's "extra-scientific origins" in global and regional geopolitics were becoming more transparent. During the delay in the completion of the project's first stage due to the tunnel collapse (roughly 1961 to 1965), the LRA dropped major components of the plan to irrigate the coastal plain of the country. Water stored behind a smaller dam after its exit from the tunnel, originally intended for Lebanon's semiarid southern region, was instead directed toward electricity generation for domestic consumption in Beirut. Only after vociferous protests from political and religious leaders in the south, who lobbied for diversion of water directly from the Karoun reservoir toward areas that were impoverished and lacked irrigation facilities, did the government agree to "remodify" the Litani project.[82] Ultimately, the government adopted the proposal to divert water to the south.[83]

The rationale for construction of the Bisri Tunnel, the severe environmental complications associated with its construction, and subsequent conflicts over how the waters of the Litani would be used all reveal how scale-making projects at regional and national levels made a shambles of the desire to focus development efforts on a basin scale. Paradoxically, the transfer of the Litani's flows outside of its "natural" basin to achieve the national goal of providing electricity for Beirut, and to respond to regional geopolitical concerns over any water resource development in proximity to Israel, led directly to an extra-basin solution and the biophysical calamities that ensued. The Bureau's engagement with the Litani River and the development of its basin ended far more quietly than seemed likely given the ambitions professed at the launching of the project in 1951. The Karoun Dam and other project components were constructed in the 1960s by a consortium of French firms, not the American engineering companies the Bureau and State Department had hoped for.[84] American geopolitical engagements in Lebanon, already waning by 1958 when the Eisenhower administration concluded that its efforts had successfully staved off Soviet inroads in the Middle East,[85] had greatly diminished by this later period. Despite the Bureau's ambitious vision of comprehensive development of the Litani basin, the network of irrigation developments and additional multipurpose projects on the Litani—and the dream of an "Arab TVA"—failed to materialize.

Conclusion

The enthusiasm and optimism that characterized the Bureau's early statements on technical assistance to promote development stand in stark contrast to the agency's actual experiences in the Litani basin. Analytically, the (geo) political strategies and technical decisions that feed technopolitical interventions should not be universalized across space and time; rather, these dynamics are influenced by very specific historical and geographical circumstances. In addition, the technopolitics that characterize the development of the Litani River basin were initiated and shaped by a variety of scale-making projects associated with processes and decisions envisioned, at different times, as national, regional, and global in scope. This observation encourages consideration of a more geographically inflected notion of technopolitics, particularly if we are attentive to recent theorizing on spatial scale.

Scale manifested itself in the technopolitics of the Litani basin in several important ways. From the perspective of US foreign policy at the onset of the Cold War, development of the Litani basin and Lebanese territory were perceived as minor parts of, first, a global geopolitical vision of a bipolar and contested world where newly independent states needed convincing of the efficacy of alliance with the United States, and second, a regional vision of the Middle East seen exclusively in terms of US strategic interest. The technopolitics that produced the Litani basin plan and the subsequent infrastructure projects was thus an outcome of global and regional scale-making projects. Yet the desire to make technological assistance serve geopolitical and developmental ends, and the scale-making projects associated with those ends, were not exclusive to US actors. In Lebanon, State Department officials and Bureau engineers encountered politicians and bureaucrats wary of US intentions, eager to achieve economic development, aware of their own regional vulnerabilities, and sensitive to the domestic political ramifications of implementing a massive development scheme. Finally, the ostensibly natural and common-sense biophysical integrity and scale of the river basin was simply contravened by the decision to divert water underneath a mountain range. Moreover, the local geological conditions near the site of the Karoun Dam and the Bisri Tunnel—which themselves constitute a unique spatial entity with a distinct "project" scale—were accorded secondary status in comparison with national and regional political aims. If we lose sight of the various scale-making projects at work within the geopolitics (and technopolitics) of development, we risk overlooking the ways that networks made up of technical experts and (geo)political calculation operate "at a distance" within marginalized locales. River basin development, especially perhaps the highly incomplete version seen in the case of the Litani, is always an expression of political power as much as an exercise in technological competence.

In that vein, it is perhaps ironic that the Litani basin development program failed to convince the Lebanese state to become an active geopolitical collaborator of the United States in any meaningful way, and that it contributed to economic development in a very limited fashion. Ultimately, development of the Litani basin became enrolled within a broader assemblage of power relations that corresponded to the global spread of large dams and river basin development. Governments and development planners of the "underdeveloped" regions perceived large dams and river basin planning as mandatory elements of a universalized approach to the exploitation of water resources. So while the Litani project may have failed in crude geopolitical and developmental terms, it was perceived within the Bureau of Reclamation as a foundational example of how to disseminate the ideal of the modern river basin and its technologies globally.

FOUR

Ethiopia, the Bureau, and Investigations of the Blue Nile

... and I remember that in the twinkling of an eye the following slogan appeared in the streets of the capital:

As soon as the work on the dams is done,

Wealth will accrue to everyone!

Let the slanderers spew their lies and shams—

They will suffer in hell for opposing our dams!

—former palace assistant to Emperor Haile Selassie, speaking of events in Addis Ababa during June–July 1974[1]

Along with the Litani River initiative, the Bureau's other major international effort of the 1950s was the Blue Nile investigation in Ethiopia. The Bureau's role in Ethiopia's water resource development strategies stretched over a nearly two-decade period from 1951 to the late 1960s. While some of the development focus fell on the Awash River basin, the Bureau's primary activities centered on irrigation and hydropower projects in the headwaters of the Blue Nile (fig. 4.1).[2] The Blue Nile program was critical for both institutional and political reasons. It helped to revive the Bureau's international program as it came under threat due to disinterest within the Bureau and lack of funding from Washington.[3] And like the Litani program, it became deeply embroiled in both the national politics of development within the host country and broader geopolitical machinations centered on proposed cooperative development of a transnational basin. Despite these parallels, the Blue Nile initiative was unique. The potential development of the Blue Nile, which forms the headwaters of a prominent international basin, was used by both the Ethiopian and American states as a lever of influence over downstream states, particularly Egypt, and helped hold together a rather tenuous geopolitical alliance. However, the technopolitical network forged

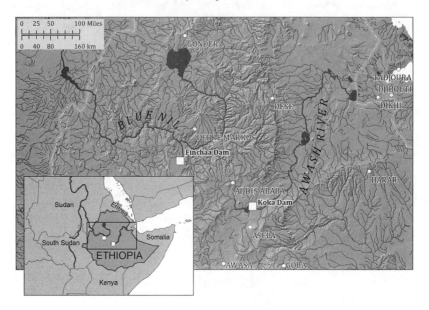

4.1. Map of Ethiopia, showing the Awash and Blue Nile River basins and sites of prominent hydroelectric dams.

within the Blue Nile investigations was constantly challenged by the vicissitudes of the Bureau's daily efforts to exercise its technical expertise in a biophysically and politically complex environment. Such challenges ranged from the logistical slogs of active field sites to, at broader scales, negotiating the complex geopolitical milieu of technical assistance. The Blue Nile case underlines the point that technopolitical networks are never complete, but are always works in progress.

I proceed with a brief interlude on the evolution of the Bureau's foreign activities during the critical period from the early 1950s to the mid-1960s, which might be considered the high point of the Bureau's foreign engagements. This discussion will further contextualize some of the dynamics of the Litani case (see chap. 3) and will set the stage for discussion of the Blue Nile program in this chapter and of the Bureau's contemporaneous experiences in the Mekong River basin in chapter 5. The remainder of the chapter looks at the Bureau's initial experiences with river basin planning in Ethiopia, some of the regional geopolitical considerations of concern to US officials, Ethiopian dissatisfactions with development of the Blue Nile, and finally, the actual outcome of the Bureau's multiyear investigations. While it is hardly surprising that wealth did not in fact "accrue to everyone" as a result of Haile Selassie's hydro-dreams, Ethiopia's early experiments with the development of hydroelectricity provide an instructive example of the ideological flexibility and staying power of large dams and basin-oriented river alteration. However, these experiments also presaged Ethiopia's current water resource development programs, reiterating the capability of technopolitical networks constructed around specific projects to transcend momentous political changes at national and global levels.

Technical Assistance, Point Four, and Liberal Capitalism

By 1959 Bureau engineers had served in Pakistan, Ceylon [Sri Lanka], Malaya [Malaysia], Nepal, Korea, Turkey, Japan, Afghanistan, Israel, Iraq, Iran, Jordan, Lebanon, Liberia, Egypt, Ethiopia, Greece, Saudi Arabia, New Zealand, Venezuela, Uruguay, Costa Rica, Brazil, Yugoslavia, the Philippines, and several other countries—a remarkable feat by any measure.[4] During the 1950s, the Bureau received roughly 1,600 engineers, economists, and other technical specialists—from 49 different countries—seeking training in water resource development. According to one study, the Bureau's Denver office was host to roughly 2,700 foreign trainees from South Asia and the Middle East alone over the period 1946–1990.[5] To the Bureau, this expansion of its international scope was confirmation of its status as the world leader in water resource development. However, it also represented a relatively rapid comprehension on the part of the US government that its technical assistance programs had political value.

As described in chapters 2 and 3, augmentation of the Bureau's activities in foreign countries was highly contingent on the political dynamics of foreign aid. Any interpretation of subsequent actions by the Bureau, of where and when it responded to requests for assistance and the type of assistance it delivered, must be cognizant of how the broader geopolitical architecture of the Cold War facilitated the transfer of technical expertise and, in turn, how the requirements of technical knowledge (e.g., detailed fieldwork, specific biophysical knowledge) constrained the political objectives of US foreign policy. As evidenced by its negotiations with the State Department in the late 1940s, the Bureau was deeply concerned over its expanding role as a technical assistance arm of the US government lest this role detract from its primary mission as a domestic resource management and development organization. Indeed, the only exception to the Bureau's general rule of never participating in international activities if such activities were interfering with its domestic program was when the "Department of State advised the Department of the Interior that a particular activity would be in the national interest as part of the United States Foreign policy." In this situation, "consideration would be given to engaging in such activity even though some adverse effect on the domestic program might occur."[6] One of the most significant things to come out of the early relations between Reclamation and the State Department was a better acquaintance of the latter with the "unique skills and abilities" of the former. This was important due to the simple fact that, as mentioned in chapter 3, the Bureau's water resource development specialists were skeptical of the capacities of State personnel to comprehend the technical and scientific nuances of their work. As the Bureau was to learn to its chagrin later, the diplomatic and foreign policy skills and aims of the State Department were equally ill understood within the engineering offices of the Bureau. These tensions would come to the fore in the Blue Nile project as well as the Mekong project. Aside from geopolitical questions, the Bureau also had to come to grips with how its "unique abilities" might assist in one of the United States' key aims in engaging with the Third World: the advancement of American economic interests via the exploitation of resources and opening up of new markets.[7]

By the time the Bureau's overseas activities were established in the 1950s, the Eisenhower administration was promoting economic assistance in a way that would, in theory, enhance the expansion of both foreign investment and, eventually, domestic investment (allowing newly independent societies to "catch up" economically) in the underdeveloped areas of the world. Concerns over the role of aid in general and the Bureau's role in technical assistance had been voiced earlier. For example, the Bureau took part in a wider debate within the Truman administration regarding the appropriate approach to economic development and foreign aid. Such discussions, from the Bureau's perspective, "were required to reconcile the divergent views that emerged as to where and how the line should be drawn between the exchange of technical knowledge and skills and the provision of capital investment." Bureau leadership quickly recognized that the "body of opinion within the [Eisenhower] Administration favored the provision of straight technical assistance as a forerunner of capital financing through private enterprise."[8] The actual construction of dams and other water infrastructure was not perceived as the Bureau's central purview. This was clearly the case when forces within the State Department balked at John Savage's vision of the Bureau designing and constructing the Yangtze Gorge project (chap. 2), and this view was also evident in deliberations over the Bureau's role in Lebanon (chap. 3). Rather, the Bureau would be a catalyst, setting the stage for construction through field reconnaissance and feasibility studies, perhaps some design work, and other preparatory kinds of labor. The actual construction of projects was to be carried out by private operations, preferably American firms. In the words of an assistant commissioner of reclamation, "American contractors and engineering firms complement this [technical assistance] program, having an expanded market abroad resulting from this Government to Government technical assistance."[9] This attitude was confirmed very early in the Bureau's experiences in Ethiopia, as we will see later in this chapter. Of course this tack also provided the Bureau—and its international programs—firmer institutional and political ground during interagency struggles within the American state. In fact, the Eisenhower administration's approach to the agency's economic and technical assistance overseas was a reflection of its stance toward the Bureau's domestic activities in the 1950s. The Eisenhower administration, concerned that the United States' "natural resources program" was characterized by "exclusive dependence on Federal Bureaucracy," curtailed the Bureau's domestic spending and advocated for a "partnership" of states, private citizens, local communities, and the federal government. The "partnership" philosophy was "strongly conservative, oriented towards free enterprise and anti-New Deal," and, combined with the political climate generated by McCarthyism that cast nearly all federal programs in a socialist light, resulted in several very lean years for the Bureau.[10]

Throughout this period and into the 1960s, Bureau leadership was careful to advertise its foreign activities in subtle ways; the agency was wary of being perceived as promoting foreign dependence on US aid and otherwise interfering with American business interests. At a speech before a group of professional engineers, Commissioner Floyd Dominy accentuated that

> the intent of Congress is clear. Private enterprise abroad is to be encouraged to the fullest extent. And this, to my estimation, is being done by the Federal agencies involved in this program. On the other hand, Congress also recognizes that Government engineers have a legitimate and necessary role to play in professional engineering across the seas. This participation is not on the basis of our doing a job more cheaply than a consulting firm, but on our ability and experience in establishment of water policy, formulation of Government operating standards and criteria, and inter-ministerial coordination and management.[11]

Dominy's references to "ability and experience" certainly invoke a legacy of prodigious technical knowledge and legitimacy in water resources research and development that echoes John Savage's work on the Yangtze (see chap. 2). Dominy was adamant that "when our technical work abroad reaches the developmental stage, the Bureau's participation ends," noting that numerous "engineering firms in the United States have obtained clients and have benefited by this catalytic process. And, of course, the reports and data on our investigations are made freely available to the profession through the sponsoring agency."[12] Moreover, Dominy and previous Bureau leaders knew very well that their expertise in technical assistance could be a formidable tool not only for securing alliances with recalcitrant governments negotiating the US-Soviet geopolitical spectrum, but for promoting national economic interests in several ways. As Dominy observed, "The significance of our service abroad is that we are contributing to the stimulation of natural resources development in many countries of the world," which "makes it possible at a later date for many United States consulting firms and construction contractors to perform work on a global scale." The Bureau's efforts, according to Dominy, thus brought an obvious advantage to American firms seeking overseas work.[13] Moreover, Dominy's invocation of "policy" and "operating standards" suggests the straightforward logic of the Bureau's creation of an administrative environment within host countries that would be highly conducive to the entry of American firms that would have prior experience with this same environment in the United States. In this fashion, the technopolitical networks that the Bureau helped create in the crucible of technical assistance simultaneously projected the Bureau's approach to water resource development as the global acme and fostered, in theory, economic opportunities for American corporations in a receptive setting.

The Bureau's activities would also support US economic and political objectives in less obvious ways. First, both the domestic training program and in-country training activities helped promote US economic interests. The training of thousands of technical personnel in the United States and "building up professional associations overseas" helped engineers and other water experts from developing countries "to understand and appreciate American methods and technology," which in turn would facilitate "opportunities for US business enterprise."[14] The Bureau's "pioneering work" in regions such as Asia and Africa, where development was initially "carried out under European influence and the work was performed under force accounts," influenced water development officials and agencies to "utilize the system generally utilized in the United States calling for the use of consulting engineers and contractors," which, of course, "results in additional work for American firms."[15] Second, the Bureau pointed to the "fraternity" of non-US engineers who would benefit American foreign policy aims. Throughout the 1960s the Foreign Activities Division listed the "distinguished" high-level personnel, many of them heads of powerful water bureaucracies, who "once studied or spent extended periods of observation with the Bureau," including officials from Thailand, India, the Philippines, Sudan, Turkey, and the United Arab

Republic [Egypt]. These men, argued the Bureau, had now become "senior policy makers who occupy positions of trust and respect," and "play a vital role" in their countries' relations with the United States:

> Without exception, they are laudatory in their praise of the United States and of the training and treatment they received here. They are memorable ambassadors of goodwill for the United States in their own countries.[16]

The Bureau's fulminations on this point raise a crucial aspect of its overseas work and speak more broadly to the lasting impacts of technical assistance of the kind centered on the "concrete revolution" in large dams and river basin development. The dominant understanding of the Bureau's foreign policy role, at least by the 1960s, was founded in its function as a catalyst for private enterprise. From the beginning of the Bureau's escalation of its global work during the Eisenhower administration, the aim of promoting American business interests seems to have taken priority over the broader and somewhat more nebulous goal of steering the attitudes and policies of newly independent states toward the US sphere of influence. Either goal would be tremendously hard to assess, and it is striking that Dominy's speech makes no specific mention of an American firm that benefited directly from Bureau activities.

There is no question that Dominy, like John Savage, deserves a great deal of credit for expanding the scope of the Bureau's overseas activities. He was certainly responsible for bringing the Bureau's international efforts greater attention, both in the public eye and within the arguably more crucial corridors of Washington, where appropriations were committed and overseas deployments were underwritten. L. W. Damours, chief of the Bureau's Foreign Activities Division, noted in 1965 that it was due to the "personal interest, leadership, and support of the Commissioner" that his office was in a "healthy condition" in terms of financial support and expanded duties.[17] Dominy lent his considerable charisma and political savvy to the Bureau's ongoing activities in Ethiopia and was instrumental in convincing the Mekong basin countries to move forward with the Pa Mong project (see chap. 5). However, Dominy's promotion of such projects on political and economic grounds raises a critical question: In its efforts to justify its overseas work on the basis of its benefit to private enterprise, did the Bureau abdicate its responsibility as a government agency to promote the public good? The problematic legacy of large dams and river basin development programs in terms of social and ecological disruption certainly casts doubt on the long-term beneficence of the Bureau's work. I return to this question in chapters

6 and 7. The more important question in the context of the 1960s is, did the work of the Bureau and its technical assistance programs substantially benefit the United States during the Cold War? Before any conclusive answer can be given, a more thorough understanding of what actually confronted Bureau experts as they encountered specific social and ecological conditions in places like rural Ethiopia is required.[18]

The Awash Basin Investigations, 1952–1954

The history of Awash River basin development and the Bureau's role in Ethiopia in the early 1950s prefigures the evolution of the more extensive technopolitical networks that were mobilized and maintained during the Bureau's Blue Nile studies in the early 1960s. Indeed, many of the obstacles that confronted Bureau experts in the Blue Nile case were paralleled in the Awash studies. The hydroelectric potential of the Awash River basin in central Ethiopia (see fig. 4.1)—a rare inland river system located in East Africa's Great Rift Valley—was first broached by the colonial Italian government, which controlled Ethiopia from 1936 to 1941.[19] While the Awash and Blue Nile studies were clustered together within the same broader investigation of water resources in Ethiopia (initiated in 1953), Bureau work on the Awash progressed more rapidly.[20] A team headed by Thomas A. Clark, named the chief planning engineer, and including six other personnel (three engineers, a soil scientist, and a hydrologist) arrived in Addis Ababa in March 1953 (fig. 4.2).[21] In general, the Bureau team focused on geological and topographical studies in the Awash basin in support of ongoing work initiated by the Ethiopian government after it had decided to proceed with construction of the Koka Dam, a hydroelectric project located approximately 80 kilometers southeast of Addis Ababa.[22] Funding for the Koka project, including a power station and transmission lines, was provided for in the Paris peace treaties of 1947 that followed the conclusion of World War II and supplied by Italy through the Reparations Fund.[23]

A series of monthly and annual reports from the Bureau team reveal that work on the Awash in these early years focused almost exclusively on the development of a stream gauge network for determining basic hydrological data. These annual reports counter the rather narrow portrait provided in State Department documents and classified communications regarding the political importance of economic assistance to Ethiopia. Additionally, the reports reveal a good deal about the everyday pitfalls of implementing river basin development and investigating a complex landscape for potential dam sites. As the actual terrain of the Awash basin was brought more and more

4.2. Bureau staff at field site in Awash basin, Ethiopia. *From left to right:* Dallas Watkins, chief of engineering surveys; Keith Davis, supervising soils scientist; and Earl Meneely, supervising hydraulic engineer. Courtesy of National Archives, Denver, Colorado.

into contact with the knowledge of Bureau water experts, the materiality of the basin delayed and complicated the construction of the technopolitical network that would come to characterize the Blue Nile studies. For instance, at the Koka Dam site on the Awash River, the research camp was situated "high above the Awash flood plain in order to avoid undue exposure to malaria." At the survey camps (fig. 4.3), the "main phase of operation" was "the procurement of food and incidental supplies," and the latter was "mostly bought 'on the hoof' and killed at the camp." The lack of equipment, the lack of transportation, and the slow pace of training the Ethiopian survey workers made the Bureau's topographical work—crucial to eventual dam construction because it would allow precise determination of the reservoir's coverage and depth—a difficult slog.[24]

Weather conditions and difficulties with measuring equipment were also problems. All surveying work at the Koka Dam site was "terminated" for a week in mid-1953 due to the "physical limits imposed by the weather." The "deterioration of the road system" leading to the dam site, which in the best of conditions was basically a "jeep track," made further surveying impossible.[25] Because surveying was "fair weather work" due to the need to gener-

4.3. Camp site for field research by Bureau of Reclamation team in Awash basin, Ethiopia. Courtesy of National Archives, Denver, Colorado.

ate drawings and maps in the field (where no adequate shelter was available), the entire crew was relocated to Addis Ababa for two months. Even the seemingly straightforward placement of an automatic recording gauge, vital in determining daily and seasonal flow rates, was challenging. Although the reasons are unclear, people on both banks of the Awash near the Bureau's investigation site "objected to the surveyors working in the area" and caused a month-long delay of the data collection.[26] An earlier communiqué noted "tribal warfare in the vicinity" of the Awash studies, which prompted the need for police protection of the Bureau's stream gauge reader.[27] Delays in getting the appropriate equipment for Bureau technical activities frequently stymied the scientific investigations. Bureau staff complained of their inability to get the necessary technical gear delivered into Ethiopia and, more importantly, to field sites. One early Bureau report notes that the Bureau program "has not developed as rapidly as originally planned" due to the long delay in shipping "some essential items of technical equipment." Bureau staff also had to serve as local ambassadors of a sort. For example, a January 1954 report mentions the need for Keith Davis and Dallas Watkins (see fig. 4.2) to drive throughout the lower Awash basin contacting provincial governors and police to alert them of imminent Bureau activities.[28]

An episode during the early stages of the Awash investigations also demonstrates the collusion of "local" politics and the need for technical information that stymied Bureau efforts. Work in the Awash, which was almost entirely focused on the development of a hydroelectric dam at Koka, could have been greatly accelerated had the Bureau been granted immediate access to the hydrological and topographical data collected by the Compagnia Nazionale Imprese Elettriche (hereafter Coniel), a consortium of Italian engineering firms that had conducted studies of the Awash's development potential during the period of Italian occupation and colonialism. In early 1953 chief planning engineer Tom Clark made a plea to US embassy officials that "we should exhaust every possibility for obtaining the data from the Coniel Company," despite a stipulation from Ethiopian officials that the Bureau staff "make a survey of the [Koka] dam site regardless of the fact that this work had already been done."[29] Coniel had been engaged in engineering studies since 1935 and had secured hydrological and topographic data for segments of the Awash Valley. It had also developed detailed designs for a masonry dam at Koka measuring roughly 20 meters high with a crest length of 180 meters, capable of storing 1.3 billion cubic meters of water.[30] Yet Coniel was recalcitrant about giving up the requested information. Clark pleaded with Ethiopian officials in mid-1954 for clarification of several issues that needed responses before the Bureau's work could proceed. The Bureau required all available stream flow data from Coniel, but had not received it. The Bureau, awaiting direction from the Ethiopian government, wondered whether it should use the "designs prepared by Coniel and have them approved by the Bureau of Reclamation in Denver," or have new designs "prepared by the Bureau of Reclamation or a firm of consulting engineers?"[31] It is not clear whether or not Coniel relented and provided the flow data.

The situation that emerged with Coniel highlights the unexpected political and logistical problems that confronted Bureau workers engaged in river basin development efforts abroad. In fact, members of the Bureau mission frequently complained about the complex bureaucratic environment that confounded what, to them, should have been straightforward technical exercises in both the Awash and the Blue Nile basins. For instance, the identity of their primary partner agency in Ethiopia was never quite clear to Bureau staff. A key Point Four official complained about the fact that the Ministry of Public Works "has no responsibility in connection with Water Resources development" even though it was (apparently) the primary institutional contact for the Bureau's biophysical investigations. Moreover, there was "no clearly placed responsibility . . . in the Ethiopian Government . . . for the collection of water resources data or for the development of planned for water conservation and utilization."[32] This ambiguity more than likely reflected conflicts within the Ethiopian bureaucracy, a political context that was seemingly far removed from what the Bureau imagined its responsibilities to be.

The Bureau's efforts in the Awash were also clearly pegged to a vision of basin-oriented water resource development, as they were in the Litani case and later in the Blue Nile basin. Tom Clark emphasized this perspective when noting the "extensive irrigation" possibilities in the Awash valley, in addition to the planned hydroelectric project at Koka. All in all, development of the Koka Dam "should consider the fullest practicable use of water, with flood control, power and irrigation making a multiple use project." Moreover, the reservoir created by the Koka impoundment would "furnish a controlled flow through the several power plants which would ultimately be built from Coca to Awash, with power water then being available for irrigation on the plains below."[33] The Bureau's vision for the Awash, and certainly for the Blue Nile as well, hinged on the creation of a modern basin wherein flowing water was captured and used in an economically optimum fashion.[34] This vision in turn depended on the collection of vital data regarding topography, land ownership, and markets for irrigated crops, in addition to laws governing water use in the region. Water resource development and the political advantages it might confer were thus contingent on technical knowledge and the quotidian methodologies that produced that knowledge. It was this intersection of technical expertise, political calculation, and biophysical knowledge production that would continuously bedevil the Bureau's efforts during nearly all of its foreign interventions.

As preordained, the decision was made by the Ethiopian government to proceed with the Koka Dam and hydroelectric power station. In his speech inaugurating the dam in 1960, Haile Selassie spoke in theological terms of the "impressive installation" and the "dynamics and fructifying powers of our water resources," concluding that it is "our bounden duty, to exploit them to the full, thereby rendering possible a surge of development in our agriculture and industry."[35] His Majesty also linked exploitation of the Awash, via the Koka project, to development of Ethiopia's other rivers, including the Blue Nile. Singling out the role of technology in attaining improvements to human well-being, he asserted, "We . . . shall never falter in our efforts to assure that the benefits of modern science and technology shall be lavished upon our people." It is likely that the rhetoric of "lavish" benefits directed toward the Ethiopian people was designed to enhance the government's legitimacy in the face of the continuing civil unrest that culminated in an attempt to overthrow the Selassie regime in December of that same year.[36] While perhaps a crude example, this speech shows how

the Koka dam, and the later development of the Blue Nile, became important elements of the discourse of development in Ethiopia and part of the
political strategy of the emperor, as the convulsions of the 1960s continued, to sustain an increasingly fragile authoritarian regime. At the time of
the Koka dedication ceremony, the Blue Nile investigations had entered a
critical phase, a culmination of years of preliminary technical studies and a
prodigious amount of political rancor involving the Bureau, the US foreign
policy apparatus, and the Ethiopian government.

Constructing a River Basin: The Bureau, Geopolitics, and Investigations of the Blue Nile

By the end of the 1950s, the Bureau presence in Ethiopia had grown to thirteen engineers, geologists, and other personnel engaged in Blue Nile investigations; these numbers had doubled by 1962, and the Bureau staff were part-
nered with roughly 150 Ethiopian staff members. The stated aim of the Blue
Nile study was straightforward: to report on and assess the potential for "the
economic development of the resources of the Blue Nile."[37] The Blue Nile,
known in Ethiopia as the Abbay (Abay) River, originates in Lake Tana, in
the central highlands of the country, and its basin covers an area of roughly
324,000 square kilometers, most of this falling within Ethiopian territory
(see fig. 4.1). The river itself flows approximately 900 kilometers from Lake
Tana before reaching the Sudanese border, where it continues to its confluence with the White Nile near Khartoum.[38] It traverses a series of deep gorges
downstream from Lake Tana as it flows through the Ethiopian Plateau. The
upper portion of the basin is characterized by savanna forest, while a mix of
woodlands, shrublands, and grasslands characterizes the more arid lower
lands near the Sudanese border. As is the case with the Litani, precipitation,
and hence river discharges, in the Ethiopian Blue Nile are highly seasonal,
with roughly 70 percent of mean annual rainfall and over 80 percent of annual discharge occurring from July to October.[39] Again in concordance with
the situation in the Litani basin, the Blue Nile basin's residents in the 1950s
were—and remain—largely rural and subsistence-oriented. According to
one Bureau report, a wide majority of the basin's 4.9 million residents "farm
small tracts of from 4 to 8 hectares, using handtools and oxen much the same
as they have for thousands of years."[40] The principal crops of the region included "grains, pulses, oilseeds, and spices" in the highlands and "sorghum,
cotton, sesame, and corn" in the more sparsely populated lowlands.[41] What
confronted the Bureau, then, was a remote basin with a population considered traditional and marginal (more peripheral than the Awash basin), one

that in the minds of Bureau engineers and Ethiopian officials alike would unequivocally benefit from the irrigation and hydropower infrastructure that comprehensive river basin development could provide.

The Bureau's original intent was to provide a program of *all-inclusive* river basin development for the Blue Nile in a pattern of technical assistance remarkably similar to the Litani and, indeed, the Awash basin cases. The original reconnaissance study, headed by Bureau engineer Tom Clark in 1952, concluded that "sufficient storage could be provided along the Blue Nile and its tributaries to permit full utilization of the river for power and irrigation in Ethiopia."[42] The Bureau's work in Ethiopia also involved the training of Ethiopian engineers in the fundamental technical skills of river basin planning (e.g., surveying, soil analysis, hydrological investigations) and, as noted in a Department of the Interior press release at the onset of the program, the "establishment of a comparable agency within the Imperial Ethiopian Government." The hope was that Bureau-trained Ethiopian engineers would someday "be in a position to carry similar work forward in other river basins in Ethiopia."[43] Additionally, as we have seen, the Bureau's activities were linked to a liberal capitalist view of foreign aid and technical assistance, one that in theory would advance US business interests and expand markets for US firms involved in dam construction and other infrastructure projects. The State Department, noting that the "United States Government has no complete engineering service such as a department of public works," argued that "it is often more advantageous to use private enterprise for engineering jobs" such as the one contemplated for the Blue Nile.[44] The Bureau was thus assigned multiple roles by the American state, but all oriented toward advancing US political and economic interests, with only partial consideration of the goals and needs of non-US governments and societies.

The importance of Ethiopia to US strategic objectives accrued largely from its position within a broader regional geopolitical context. The Horn of Africa was deemed a region of geopolitical interest as early as the 1950s by the Eisenhower administration and successive US regimes not only because of its proximity to critical routes for transporting oil, but also because of the complex interplay of postcolonial political dynamics and emergent socialist-oriented social movements characteristic of many African nations of the 1960s. Ethiopia's ostensibly modernizing leader, Emperor Haile Selassie, saw US aid as a means to both bring economic development and industrialization to his largely rural nation and, via military assistance (arms and training), bolster his legitimacy and security internally and within a politically unstable region.[45] Despite America's "good relations" with the Ethiopian government during the 1950s, by 1957 concerns had arisen about the Selassie regime's "penchant for seeking special treatment" in connection with military and economic aid.[46] In the previous year, the chair of Eisenhower's Joint Chiefs of Staff concluded that one of the actions by which the United States could "best maintain her present friendly relationship with Ethiopia" would be to ensure economic aid of a type that would "make an immediate demonstration of gains to be expected from alignment with Western countries," as opposed to the current program of technical assistance, which had "not shown any tangible results." The State Department believed that the Blue Nile studies would conclusively demonstrate these "tangible results,"[47] but their actual influence would eventually become a source of friction in US-Ethiopia relations rather than a demonstration of a "friendly relationship." Early cost estimates for a comprehensive basin development scheme, which would take ten years to come to fruition, were placed at roughly US$150 million (wherein construction costs were placed "at $25,000,000 per dam, for say 5 dams").[48] While this was certainly a very tentative estimate, it indicates the relative geopolitical importance that some in the State Department ascribed to Ethiopia and the region more broadly.

Indeed, official US concern over regional geopolitics extended to the sharing of the international waters of the Nile River basin. As early as 1952 the State Department was aware of the sensitivity of any technical investigation of the Blue Nile's development potential, given its upstream location and importance to flow rates downstream in the Egyptian portion of the drainage area. The State Department clearly understood that any cooperative development of the Nile basin was contingent on whatever "stream flow data and such other engineering information" from the Blue Nile might be required to have an "intelligent discussion" about sharing international waters.[49] Later in the decade, State Department officials posited that hydroelectric development of the Blue Nile, while largely feasible as confirmed by Bureau studies, was of increasingly grave concern to Egypt because of the potential negative impacts downstream in the Nile River's critical delta region. The Eisenhower administration was also aware of Ethiopia's ongoing "irritation" with Egypt over the latter's lack of consultation on the Aswan High Dam.[50] By late 1958 advisers in the Eisenhower administration saw conflicts over the development of the entire Nile River basin as a key component of their strategy to exercise influence over Egypt's President Gamal Abdel Nasser, who had navigated a more or less neutral path between the two Cold War superpowers' desires for influence in the Middle East.[51] For example, a National Security Council member recommended that the State Department push forward with hydrological investigations on the Nile's headwaters because this approach was "the best hold this country could possibly have on Nasser for bargaining purposes."[52] US support for river basin studies in Ethiopia thus served as a tacit reminder, and even a threat, that development of the Blue Nile would alter downstream flows.[53] This strategy led directly to the more intensive Bureau focus on the Blue Nile basin, which began in earnest in 1959 and lasted until publication of the Bureau's report in 1964. What happened over this duration is illustrative of how the technopolitical networks constituted through water resource development come together and evolve.

In 1961 over two dozen Americans affiliated with the Bureau were working with a multitude of Ethiopian personnel to determine the most appropriate sites for hydroelectric and irrigation facilities in the Blue Nile basin, and this level of involvement continued until publication of the official study report in 1964. The total cost of the Bureau investigations, funded entirely by the US government, was US$4.5 million. The final report detailed an incredible array of projects that, if built, would have radically transformed the Blue Nile into a modern basin, replete with irrigation canals, hydroelectric plants, substations, transmission lines, flood mitigation structures, and nearly complete control over the flows of the river and its tributaries (including multiple schemes on the rivers flowing into Lake Tana). Within this vision of total basin control, the Bureau team proposed 11 dams exclusively for power production (including four massive schemes on the Blue Nile's main channel), 14 irrigation projects, and 8 multipurpose projects.[54] These 33 schemes would impound over 118,000 million cubic meters of water, provide irrigation water to over 430,000 hectares of land, and generate almost 7,000 megawatts of electricity. The total cost of this proposed transformation of the Blue Nile basin was estimated at US$3.2 billion, and the full program would probably take decades to implement.[55] The imperial Ethiopian government thus had a clear program of river basin development to follow, yet one that was clearly impossible to fulfill given available resources. While the Bureau provided the Ethiopian government with a blueprint for its own TVA-style basin development, what that government, or at least its leader Haile Selassie, most genuinely desired was a physical manifestation of water control; in other words, an actual dam. The Bureau's Blue Nile studies culminated in its 1964 report, but behind that achievement lay a foundational period encompassing over ten years of bureaucratic explosions, geopolitical maneuverings, and nearly insurmountable technical difficulties.

As in the Awash studies, the Bureau was asked to perform activities and confront local conditions in way that both enveloped and went far beyond technical assistance. Undertaking preparatory field visits in advance of actual studies in the region, chief engineer Tom Clark discovered that a previous

British report on the hydrology of Lake Tana, source of the Blue Nile, "may not be as accurate as has been considered" and urged that it be "checked very carefully before using." Clark also tracked local lodging opportunities and commented on the likelihood that the Bureau team would or would not be assisted by local dignitaries. Clark assessed local political connections, noting that one provincial governor was a "son-in-law of the Emperor" and that "surveys and irrigation in this province would receive his backing." A liaison working with the Ministry of Public Works, however, was "unreliable and almost antagonistic." In Gojjan Province, the people were "Amharas and have not accepted the government of Ethiopia," and Clark's reconnaissance team had "four armed guards" ready at all times for potential mischief. Although Clark identified an area of the Blue Nile in this province "very suitable for early development of irrigation and power," the province's problems—including an inexperienced governor, bedbug-infested hotels, and roads that were "worse than terrible"—dictated a cautious approach to the beginning of survey work.[56] Yet these observations also demonstrate the complexity of the technological endeavors the Bureau was asked to perform.

Clark's initial assessment of the Blue Nile region hints at a common yet often overlooked theme characterizing many of the Bureau's overseas encounters. Despite the obvious interest in large-scale dams and irrigation works on the part of both national governments and Bureau engineers, there existed a strain of participatory water resource development that crops up in Bureau reports in Ethiopia, the Lebanon work, and numerous other cases. Such a perspective transcended narrow technical concerns. It was also commensurate with the Bureau's experiences in poor rural communities in the American West, where on-site engineers and project leaders often had to assume organizing and negotiating roles that transcended their technical expertise.[57] In the Blue Nile situation, following replacement of the "antagonistic" liaison, Clark discovered it might be a good idea to initiate conversations with "community heads in isolated communities" and to make contacts near work sites "well in advance" so that all interested parties, especially priests (given that the "church is the strongest element" in these parts), were aware of the Bureau's activities. The new liaison, Ato Asefa Temteme, suggested that local residents could help support the digging of collective wells, thus demonstrating the importance of community-based water improvements. Still, Clark described this contact as "not brilliant" and without any "administrative ability whatsoever." This odd mixture of respect and arrogance, which characterized nearly all of the Bureau's local encounters, underscores how water expertise was often stretched and distorted in its actual application.

From the outset of the Bureau's Blue Nile survey, it was clear that the Ethiopian government had a different set of expectations than the Bureau regarding the character of its technical assistance program. The Bureau—at least in its early years of technical assistance—perceived river basin development as a comprehensive process encompassing hydroelectric dams and irrigation development along with other economic and social goals. The Ethiopian government, on the other hand, wanted projects developed as soon as feasible, in large measure due to the "display value" of large dams, the power and prestige associated with modern dam technologies, and the legitimacy they would confer on their regime.[58] The Ethiopian government, and its leader Haile Selassie in particular, was adept at using the Bureau's technical assistance program as a lever to influence American foreign policy directed toward the country. A 1956 analysis for the Joint Chiefs of Staff asserts that "Ethiopian dissatisfaction with current U.S. aid . . . derives from the Emperor's belief that the loyalty his country has shown to the United States has been badly rewarded in comparison with the open-handed assistance rendered to some countries of the 'Neutralist Bloc.'" Selassie, the memorandum continues, "has indicated that he will turn to the USSR for help if he cannot satisfy his requirements from U.S. sources."[59]

Tensions between US and Ethiopian motivations for water resource development were also evident at the moment in the late 1950s when the Bureau was poised to increase its technical assistance to the Ethiopian government. Commissioner Floyd Dominy's visit in late 1959 prompted reassessment on the part of both the Bureau and the State Department—represented at this point by the International Cooperation Administration (ICA), precursor to the US Agency for International Development (USAID)—of the Blue Nile project's progress and its institutional and political trials. As part of his preparation for the official visit, Dominy received briefing documents from the Foreign Activities Office highlighting critical issues for discussion with State personnel concerning the Bureau's role in Ethiopia. These issues boiled down to, first, an extension of the Bureau's activities to include aerial mapping of the basin and, second, ongoing administrative conflicts involving (at various levels) Bureau staff, Ethiopian officials, and US personnel affiliated with the State Department. For example, the head of the Bureau's Foreign Activities Office, L. W. Damours, noted that "there appears to be an attitude by some of the ICA subordinate personnel that the Reclamation team is manpower or hired help being employed under more or less detailed direction of ICA." "Our view," the memo continued, is that "the Reclamation team there is performing an assignment under the direction of the Commissioner" and would present reports to ICA and the Ethiopian government upon completion of the Bureau study.[60]

Beyond such anxieties, the Bureau was dissatisfied over the administration of funding for the project. In a pointed letter to the secretary of the interior following his visit, Dominy raised an issue "that has been impeding the progress of the project"; namely, the handling of the joint fund co-managed by Ethiopian and American (embassy) offices.[61] In effect, the Bureau was stymied in its technical work, since it was responsible for "practically all procurement" for the project—"including such items as camping supplies, fuel, air transport, vehicles and major equipment"—and the Ethiopian co-manager of the joint fund was "insisting upon even closer control of procurement and property management" via the fund. As expressed by the project engineer (Donald Barnes), the Ethiopian co-manager's "position in pre-auditing the project" was "too deeply entrenched" to be changed since he "*regards himself as the management center* [emphasis in original]."[62] Dominy concluded that the procurement process and management of the joint fund were placing "an insupportable burden" on the Bureau's work, and that it was up to the ICA to remedy the situation, or the Bureau might be forced to "re-examine" its responsibilities under the Blue Nile initiative. This only slightly veiled threat, in addition to underscoring the commissioner's political savvy, reinforces an understanding of the Bureau's overseas operations as far more than purely technical assistance.

The question of mapping and topographical surveying was perhaps unexpectedly controversial. In 1959, as the Bureau was assessing the data it had already collected regarding the Blue Nile—which included hydrological information from stream gauges, meteorological records, records of soil types, geological data, and topographical reconnaissance survey records—a question arose over the scope of the topographical information necessary to move forward the Blue Nile studies in a comprehensive fashion. The United States Coastal and Geodetic Survey team assigned to assist with mapping of water resources in Ethiopia as part of the larger aid mission argued that more extensive mapping of the Blue Nile was necessary in order for the Bureau's final report to be comprehensive, and the Bureau's project engineer largely agreed. The Ethiopian government saw the additional mapping as having "great value to other agencies and segments" of its bureaucracy, and it perceived national security dividends if this mapping in any way proved important in assessing the Blue Nile's water resources in the face of downstream developments on the Nile. However, the additional cartographic surveys would be quite expensive, and ICA was unwilling to fund this addition to the investigations. The Bureau perceived the ICA's decision as largely "political and financial," since it would be up to ICA staff and their embassy affiliates whether or not the United States could offer substantial financial support for the project.[63] Indeed, internal memoranda make it clear that some US embassy officials wanted to limit the Bureau's role in Ethiopia to the Blue Nile work and not allow it to become overly involved in assisting the development of Ethiopia's nascent Water Resource Department. The embassy-affiliated US Operations Mission (USOM) saw the mapping work as excessive and wanted the Bureau to focus instead on a "phased, more limited and selective survey" that would allow for the selection of "perhaps five or six areas in the Basin with the greatest potential hydroelectric power production, irrigation and secondarily water storage."[64] Mapping, reasoned the USOM, would be necessary only for this more narrowly defined region. The Bureau had assumed all along that more comprehensive mapping of the basin would be forthcoming, since it would enable the production of a feasibility report (rather than a simple reconnaissance report) that could be used to identify specific projects with certainty. Yet there were other concerns somewhat hidden in these discussions. The political calculus of the deliberations over mapping was candid.

> Political factors play a dominant role in the entire U.S. assistance program to Ethiopia. U.S. acceptance of the IEG [imperial Ethiopian government] request to obtain data on which to stake a claim to Blue Nile water use which would hold up in international negotiations was based on political considerations. It is the conclusion of USOM that the undertaking of mapping in connection with the Blue Nile survey is necessary to meet the IEG objective in undertaking the project and to fill fully the U.S. commitment. . . . The undertaking of mapping also is necessary if the final report is to reflect credit on the U.S. on technical grounds. To do a job which would reflect unfavorably on the U.S. from a technical standpoint would be damaging to the U.S. prestige in other African countries.[65]

The political importance of mapping, and of the Blue Nile studies more generally, is underscored by the fact that the basin presented virtually no "immediate prospect for extensive development," especially in comparison with the Awash basin's far greater potential. These "regional considerations," argued embassy officials, warranted further State Department support for the Blue Nile study and an expanded mapping program. Only if the Ethiopian government balked at this approach should Washington be asked for additional funds. The debate went on well in to the following year, with the end result that the additional mapping was paid for via the joint fund.

Another of the nagging issues throughout the Blue Nile investigations concerned the relationship between Bureau experts and their counterparts in the Ethiopian bureaucracy. Some ICA personnel thought that Bureau staff should be more tightly under the management of USOM and the ICA officials appointed under the Point Four program. In part, these staff members felt that the Bureau engineers failed to "recognize differences between operations as performed in the US and Ethiopia," lacked "sufficient experience in the field of training natives of less developed countries," and were "performing" their duties "primarily . . . for the state-side audience rather than for the local welfare."[66] Part of their concern stemmed from a disagreement over the status of the Ethiopian engineers who were to be trained by Bureau experts in all aspects—technical and administrative—of water resource development. Their status had been a sensitive issue from the beginning of Bureau activities in the early 1950s. Ethiopian officials were anxious that their own experts receive sufficient training from the Bureau to be able to expand the country's embryonic water resource development program to additional basins.[67] Moreover, part of the Bureau's agreement with ICA and the IEG was to offer training and advice on running a water resource bureaucracy so that in some sense, responsibilities for future dam design and river basin development could be transferred to Ethiopian specialists. During the early stages of this cooperative arrangement, Ethiopia's Ministry of Foreign Affairs clearly stated that, with respect to the Blue Nile studies, "it is expected that this will be an Ethiopian survey, taking full cognizance of all the uses likely to be developed" in the Blue Nile basin.[68] However, Bureau personnel perceived their Ethiopian counterparts as inadequately prepared to be trained in dam design and other important engineering activities. Conversely, Ethiopian personnel assigned to the project balked at being pegged as "assistants" or "understudies" that had no say in project decisions.[69]

This internal squabbling reached a point in October 1960 that prompted a "very disturbing reaction" from the ICA co-director of the joint fund, who indicated to the project engineer that he was highly annoyed with the "constant bickering" among Bureau, ICA, and Ethiopian staff and was "fed up with the entire Water Studies Project."[70] In effect, the Bureau's water resource development expertise and its knowledge about how to carry out technical studies were being challenged by, on one hand, in-country State Department officials who wanted to ensure cooperative relations with politically well-positioned Ethiopian bureaucrats to enhance geopolitical ties and, on the other, representatives of Ethiopian agencies who wanted greater control—financial and managerial—over the water resource studies being carried out in the Blue Nile region. In the meantime, what had happened to the vision of large dams and river basin development that animated both the Bureau's work and the Ethiopian government's development goals?

From Basin to Dam: The "Projectizing"
of Water Resource Development

Initially, it is perhaps difficult to grasp how the need for armed protection of stream monitoring equipment, frustrations over personnel, conflicts over the procurement of equipment and vehicles, and controversies over the cost of mapping could exert such a substantial influence over the origins and construction of large dams. A partial answer can be found in the intricate relationships and connections built over time within the technopolitical network represented by development of the Blue Nile and the proposed construction of water resource infrastructure. Before its actual construction, a dam's foundation consists of tangled sets of networks involving state-centered ideologies of development, technical expertise, and—in the case of the dams fostered through the Bureau's work—contentious and constantly evolving geopolitical relationships. And yet a dam as a development project sustains the networks that enable its construction in the face of political and economic obstacles in a way that is more difficult for other kinds of development projects. It is difficult to imagine the principal actors in the Blue Nile case—Bureau experts, US embassy officials, Ethiopian bureaucrats, and the emperor himself—persisting in their advocacy for a development project designed around, say, improved agricultural techniques in the face of hundreds of frustrations and geopolitical maneuvers. A large dam concretizes (materially) the diverse networks that might otherwise be employed in the service of smaller, more dispersed development interventions (e.g., the community-based well projects hinted at in Bureau reports) that might be more effective in meeting the immediate needs of the rural populace.[71] As referenced in chapter 1, it was the sheer grandeur of large dams, and their promise of ample power to hasten industrialization while simultaneously conferring irrigation and flood control benefits, that captured the imagination of states throughout the developing world. These imagined benefits were bolstered and heightened as dams and river basin development became embedded via technical and economic assistance programs within the foreign policy approaches of more powerful global actors. In the Blue Nile, this constellation of networks eventually zeroed in on one rather mundane project, the Finchaa Dam.

In the mid-1950s, as noted earlier, Haile Selassie had expressed his frustration over the lack of substantial US military and economic assistance to Ethiopia, which was, according to him, more deserving of aid than neutralist countries such as India. Additionally, the emperor was "deeply hurt" and "incensed" by the announcement in December 1955 of US support for the Egyptian regime of Gamal Abdel Nasser by Secretary of State John Foster

Dulles, given the Nile's importance to Ethiopia and the slow progress of the dam building agenda within his country.[72] Efforts on the part of the United States to mollify the emperor by increasing general levels of military, economic, and technical assistance (in which the Bureau's activities were obviously included) to Ethiopia were relatively successful, yet the Blue Nile studies did not proceed as rapidly as hoped. The United States, as in the case of Lebanon, was lukewarm concerning the strategic importance of Ethiopia to broader American geopolitical goals. By 1959 Haile Selassie was again discouraged with the pace of the Blue Nile studies. Ethiopian officials consistently pointed out to embassy and Bureau staff that the emperor "desired to construct an economic water use project in the basin as soon as it was economically practical."[73] The Bureau and water resource staff in the embassy countered that any substantive reconnaissance that served to "wisely plan" development of the basin would require at least seven years, and more likely even eight to ten years, given the challenges of water resource investigations in Ethiopia.[74] The IEG apparently wanted a "'quickie' type of survey" that would rapidly identify a site for hydropower development and was willing to employ a private firm for this operation if the Bureau and the US government were unwilling to undertake it.[75]

These types of considerations led to a remarkable admission by the USOM program officer for the water resources project. Given that the main approach of the Bureau up to this point was to "move slowly and to accept the necessity for thorough, long-term analysis of water flow, the geological features of sites and soil characteristics," an acceleration of the studies would amount to a "projectizing" of the Blue Nile survey to achieve political objectives, since an economic rationale for rapid hydropower development on the river was lacking. Moreover, according to this line of argument, Ethiopia was not "sufficiently important politically to the United States" to warrant an acceleration of this type. The program officer also warned against pushing the undertaking of surveys to such an extent that "the IEG would feel we were . . . committed to assist in financing" of projects. This communication, drafted in preparation for a discussion with Bureau of Reclamation officials, also offers rare insight into how ICA and State Department personnel attempted to "manage" Bureau expertise and operations. The program officer identifies the "major problem in dealing with the BuRec [as] . . . that of drawing the line between determining the economic potential of the Blue Nile resources and participating in the development of that potential."[76] In other words, the Bureau was asked to determine the engineering, hydrological, and biophysical requirements for river basin development, but it should be left to the Ethiopian government and other foreign governments to supply the means of financing such development. As much as this stance reflects a keen desire to separate technical expertise from broader geopolitical processes, it should be apparent that such a separation was impossible.

By 1964 the Ethiopian government's frustrations with the perceived recalcitrance on the part of the United States to provide funding for a hydroelectric dam on the Blue Nile, despite completion of the Bureau's report, had reached a boiling point. In a series of interviews with Washington-based journalist Drew Pearson, Selassie spoke to this frustration:

> We . . . asked the Americans to build a dam on the Finchaa River, a tributary to the Blue Nile. It took them five years to make a survey and the dam is not even started. Meanwhile, you have doubtless heard of the big fuss further down the Nile where the Russians dedicated a much bigger dam which they built not in 40 years but in about five years.[77]

Furthermore, Pearson ominously observed, "not only Ethiopia but all Africa has watched the Russians complete the first phase of a far greater dam in about the time it took the United States to make a survey." During the same interview Selassie diplomatically pointed out, "I do not like to belittle the aid given to Ethiopia by the United States. It has been helping to a certain extent. But if, for instance, the United States had built the long-discussed dam at Lake Tana, it would have done for us what the Russians have done for Egypt at Aswan."[78]

These articles instigated a prompt discussion within the State Department. The US ambassador in Addis Ababa sent a communiqué to Washington asserting that Pearson's articles were "riddled with major and minor errors of fact and interpretation." Noting that the Blue Nile basin project "was for a comprehensive renaissance of the area, not for building a dam," the ambassador forcefully argued that the "identification of potential dam sites was incidental to the much larger objective." Moreover, the Finchaa Dam "was requested nine months ago (October 1963), not five years as Pearson quotes the emperor as saying."[79] The official reaction to the article's claims merits attention for several reasons. First, those claims demonstrated that the dissemination of large dams was tangibly linked to geopolitical strategies, both on the part of the purveyors of dams as technical assistance (i.e., the US government) and the recipients of dams as technical assistance (i.e., the Ethiopian government). Yet the links were conceived in very different ways by the actors involved. This example also highlights the messy connections between river basin development as a comprehensive approach and large dams as a key component of that approach. Haile Selassie surely saw large dams, and the Finchaa project in particular, as the key objective of his agreement to allow Bureau engineers to conduct surveys of the Blue Nile.

In part due to the attention prompted by the emperor's comments, the Finchaa Dam emerged as the focus of US support following completion of the Bureau's studies in 1964, and various agencies lobbied the Johnson administration for a US$21.7 million loan to help bring it to completion. In a memorandum to President Lyndon Baines Johnson in 1966, the director of the Bureau of the Budget referenced fears of growing Soviet involvement in the Horn of Africa and argued that the United States "has a serious political stake in Ethiopia as a moderately pro-western state with a good record of political stability and considerable influence in the African world." Finchaa, asserted the director, is in "both economic and political terms . . . a worthwhile project."[80] David Bell, the head of USAID and a close adviser of Johnson's, also sang the praises of the Finchaa project. The Bureau's 1964 study and subsequent feasibility study of the scheme, Bell pointed out, "concluded that the project as conceived was technically and financially sound, and that it would make a significant contribution to the economic development of Ethiopia." For Bell, the political advantages of funding Finchaa were obvious: "It is in the U.S. interest to ensure . . . the continuance of a moderate pro-western government with considerable influence in African affairs." Furthermore, Bell warned that "Ethiopia's continuing leadership within Africa can be assured only if a reasonable rate of economic progress is evident internally."[81]

Ultimately, US lobbying efforts paid off, and the World Bank approved a loan of roughly US$21 million for construction of the Finchaa Dam in 1969. The dam was completed in 1974. Of the dozens of hydroelectric and irrigation projects proposed by the Bureau experts in the original plan, it was the only major project to be implemented. As noted earlier, the Bureau conceived of projects such as Finchaa as benefiting the United States by creating financial opportunities for US engineering firms, and USAID and the Ethiopian government agreed to move ahead with a plan that resulted in six American companies submitting proposals to carry out more detailed investigations in advance of the dam's construction.[82] By that time the Bureau team had largely divested itself of its role as coordinator and adviser for river basin development in Ethiopia. In addition, the State Department's strategic interest in the Ethiopian government was waning by the mid-1960s. In a fashion eerily similar to the case of the Litani River basin and eventual construction of the Karoun Dam, the United States largely failed to reap any financial benefits from its long history of technical assistance. In terms of the supposed geopolitical benefits of technical assistance, it is worth highlighting the assessment of an American who had lived in Ethiopia for several years, who observed in 1967 that while "America is more deeply involved in Ethiopia than in any other country," and while its economic assistance programs were perceived to be "generally beneficial," American influence in the country had "provoked hostility among all groups in the population" due to the presence of the massive US-operated Kagnew Station military communications installation.[83] New geopolitical currents swept through Ethiopia and East Africa in the 1970s as Haile Selassie was overthrown and successive socialist regimes with strong links to the Soviet Union assumed control. By this time as well, the Bureau's excursions into international development and technology transfer had dwindled to only a few scattered programs.

Conclusion

The Blue Nile investigations were not the pathway to Ethiopia's wealth and development, as envisioned in its emperor's slogans of the early 1970s. They do, however, underscore the evolution of a technopolitical network that was mobilized in the service of dams and river basin development. This network—unique to the Blue Nile yet reflective of similar assemblages generated through Bureau activities throughout the tricontinental world—was a complex amalgam of technical expertise, local political ecologies, and geopolitical inspirations. Despite the specific characteristics of the technopolitics of the Blue Nile interventions, one of the more notable aspects of the United States' use of large dams and river basin planning as strategic tools is the remarkable similarity of the programs of technical assistance to various places deemed "of interest" geopolitically throughout the heyday of the internationalization of large dams and river basin planning (running roughly from 1950 to 1975). In contrast, US foreign policies with regard to the different regions of the tricontinental world shifted geographically during this same period depending on what were defined as strategic US interests in different regions (i.e., the Middle East, East Africa, and Southeast Asia were seen in quite different terms).[84] In spite of these different regional approaches and changing strategies over time, technological assistance as provided via programs of water resource development remained remarkably constant in its broad contours and was viewed by proponents as universally applicable regardless of geographical contingencies. While the imagined geography of a "Third World," problematic in its own right, changed in response to perceived American interests, the image of the river basin, and the water resource expertise required to render it usable, remained remarkably constant.

Indeed, the promoters of water resource development as a geopolitical "weapon" had a tendency to reduce one river basin—whether the Tennessee, the Litani, or the Blue Nile—to *all* river basins. At first, the Bureau experts assigned to Ethiopia shared a similar perspective, and they treated the material worlds of the Awash and Blue Nile basins as largely passive and eminently transformable entities. However, as the actual terrains and peoples of the basins were brought more and more into contact with the knowledge of Bureau water experts, the materiality and political complexity of these landscapes came into clearer focus. Transformation of the Blue Nile basin was delayed in part by the torturous paths through bureaucratic, logistical, and environmental settings that technical expertise had to navigate. The conveyors of technical expertise increasingly conceded that knowledge of place matters a great deal, but this information, and any alteration of development approaches it might have suggested (recall the suggestion in the Awash case to develop linkages to community needs), almost never traveled to centers of decision making, either in Addis Ababa or Washington. This problem in part explains the failure of the Blue Nile studies to achieve any meaningful developmental goals other than the modest Finchaa Dam, which was supported as a component of American geopolitical strategies almost as an afterthought. However, the Bureau's Blue Nile investigations did help construct a rather hardy technopolitical network revolving around large dams and river basin development that persists to this day and, as chapter 6 stresses, is a highly operational element of the Blue Nile's current transformation.

FIVE

Cold War Geopolitics, Technical Expertise, and the Mekong Project

This is all part of the Vietnam thing, they [the State Department] were trying to do anything they could, you know, to show the United States' presence in Laos, and Thailand, and . . . that's the way it was. I insisted that we limit our [the Bureau's] participation to making this study. And, of course, I made it look good. I took a man from [Thailand], and a man from Laos, and actually went up the river in a small boat, and looked at the site on the ground. And I flew over it in a small plane, and made a big show out of it; then, we made a study. And it would be a great dam.[1]

—Floyd Dominy, commissioner of the Bureau of Reclamation, 1959–1969

This chapter tells the story of what might be the greatest dam never built: the Pa Mong dam. Seen by its proponents as the linchpin for development of the entire Mekong River basin, this single massive scheme consumed the energies of a succession of Bureau experts throughout the 1960s. It also became the fulcrum of US State Department efforts to redefine its engagement in Southeast Asia at a period of violent struggle during the Cold War's most intense and drawn-out proxy conflict. The case of Pa Mong highlights how geopolitical framings of the Third World, particularly mainland Southeast Asia, by the US foreign policy apparatus in the 1950s and 1960s both internalized and reinterpreted the technical expertise associated with the manipulation of flowing water through the construction of large hydroelectric dams. Despite years of study costing millions of dollars and the production of hundreds of plans and reports, Pa Mong never materialized. Still, I argue, it clearly demonstrates how particular technological things and associated forms of knowledge are both integrated within networks of political calculation and generate broader networks of expertise, ecological relations, and in this case, geopolitics. These broader technopolitical networks often persist as active participants in institution building and national economic development strategies far beyond their original mandate of water resource development. How did the changing geopolitical conditions in the Mekong engage with and alter the technical expertise designed to generate large-scale water resource development? In turn, how did the geopolitical visions adopted by the US state help create new geographies of development—particularly in the Mekong context—via the diffusion of large dams and the technical expertise that accompanied them? This chapter traces the history of a project that firmly established the circulation of technology and politics that first imagined and eventually co-produced the Mekong basin as a "resource" in need of exploitation. The dam's nonexistence is a moot point; the technopolitical networks it helped assemble remain lively participants in ongoing debates over the transformation of the Mekong River basin.[2]

In 1961, a scant two years after becoming commissioner of the US Bureau of Reclamation, the redoubtable Floyd Dominy received a request from the US State Department that he personally travel to Southeast Asia for an important meeting regarding the proposed construction of a massive dam on the Mekong River. After flying to Bangkok, he was met by the American ambassador to Thailand, who informed the commissioner that he was authorized by the US government to "tell the Prime Minister [of Thailand] that the United States will help underwrite the construction of the Pa Mong Dam on the Mekong River." Dominy's garrulous response, recounted years later, is revealing:

> Mr. Ambassador, as Commissioner of Reclamation, I can't support any such nonsense. . . . Nobody knows anything about that damsite. All they know is that there's a canyon there that's narrow [and] a big river. . . . Nobody knows the hydrology, nobody knows how many villages you'd flood out with the reservoir, nobody knows what the geology is, whether we can actually put a dam in that canyon safely or not. None of these things are known, so the most you'll get me to say on behalf of the United States Government is that we'll make the necessary studies to see if a dam in the Pa Mong dam site area is feasible.[3]

Dominy recognized the need to "make a big show out of it," referencing the desire to demonstrate America's technological prowess when it came to developing water resources. But who was this show for? And what kinds of knowledge and expertise did this show require?

As we have seen in previous chapters, the *idea* of river basin development including massive dam projects was extremely attractive to State Department officials as a demonstration of American technical assistance and financial largesse to newly independent states. However, Dominy's remembrances clearly show that the expertise required to undertake such ambitious schemes rested on a detailed understanding of river hydrology and geomorphology, geological conditions at the dam site and in areas slated for irrigated agriculture, and a host of other biophysical and socioeconomic characteristics of the basin—in short, all the knowledge necessary to undertake comprehensive water development. This expertise, as in the case of the Yangtze Gorge project and the Litani and Blue Nile basin schemes, came into conflict with equally critical forms of calculation: the geopolitical goals of the American state and the developmental goals of the nation-states sharing the Lower Mekong basin (Thailand, Laos, Vietnam, and Cambodia). These aims revolved around the provision of economic and technical assistance to "underdeveloped regions" as a critical component of a broader strategy to simultaneously contain the political, economic, and ideological influence of the Soviet Union (and China) in Southeast Asia.

Of its numerous international activities, the Bureau's work in the Mekong basin and on studies specifically related to construction of the Pa Mong scheme constitutes the most intensive overseas activity in its history.[4] The Mekong project represented a degree of difficulty—technological *and* geopolitical—far more challenging than other Bureau endeavors. The sheer scale of the proposed dams, the Mekong basin itself, and the programs that would develop its resources were unlike anything the Bureau had contemplated (or ever would). The human and nonhuman actors enrolled in what would become the Mekong's technopolitical network were, and remain, multiple and diverse: the flowing water of a river basin that encompasses six sovereign states (China and Burma in the upper reaches, and the Lao People's Democratic Republic, Vietnam, Thailand, and Cambodia in the lower basin); the changing officials and regimes of those states; a diversity of fish species that constitute the most productive inland capture fisheries on the planet; the millions of rural inhabitants of the basin whose livelihoods are primarily or partially dependent on those fisheries; the international and intergovernmental organizations that have in part governed Mekong development activities; the assortment of intra- and extra-basin advocacy organizations critical of Mekong governance; the hundreds of engineering and consulting firms engaged in infrastructure development; the array of scientists investigating the basin's complex biophysical dynamics; and many more. At the time of the Bureau's entrée into the Mekong in the mid-1950s, this density of actors was muted, but tracing the evolution of Mekong development and its geopolitical underpinnings offers a glimpse into how technopolitical networks are constructed and become increasingly convoluted.

As in other chapters, I highlight the tensions between the actual work of the engineers engaged in river basin studies and the broader geopolitical aims of policy makers in the State Department. These tensions were not simply indicative of bureaucratic "turf battles" or organizational dysfunction, but reflected a deeper struggle between the perceived value of using technical assistance as a geopolitical tool, as professed by US foreign policy experts, and the approach of Bureau of Reclamation engineers in their desire to measure, calibrate, estimate, and otherwise examine the necessary biophysical details of a dam site—to generate multiple "inscriptions," in Latour's sense[5]—before moving ahead with a project. Yet Pa Mong reflected an additional tension, more internal to the Bureau, over how it applied its expertise to the Mekong River as a developmental challenge. Documents emphasize repeatedly that Pa Mong was simply the beginning of a much more ambitious program of river basin development. It was to be, in the words of a promotional document produced by Bureau officials, the "sinew of development" that would transform the entire basin through the construction of multiple hydroelectric projects, irrigation works, advanced agricultural production, agro-industrial development, and other accouterments of the modern river basin (see chap. 3). This imaginative geography of a transformed Mekong basin was propagated, initially, by US State Department officials convinced of the efficacy of water resource development as a geopolitical tool. The Bureau experts working on Pa Mong were hesitant at first to embrace this representation in the face of significant technical and economic hurdles. Eventually, however, they adopted the vision of basin-scale development that came to be known as the "Mekong project." Bureau expertise thus played a key role in co-producing and legitimizing an imaginative Mekong geography (Dominy's "big show"), which in turn generated a legible pathway for efforts (in later decades) by the riparian states and affiliated development agencies to materially alter the basin's biophysical processes through large-scale water resource development.[6]

Remarkably, the actual biophysical conditions of the Pa Mong dam site and the larger Mekong basin, and the sociocultural and political-economic networks within which this massive water development scheme was embedded, were inconsequential until planners realized that these conditions and networks presented significant obstacles to the continued application of either technological know-how or geopolitical strategizing. While Bureau engineers and bureaucrats were able to effectively negotiate domestic politics and material conditions as they reworked the geography of the American West, doing so in the Mekong basin proved more difficult. Still, the creation of an imagined Mekong geography—founded on basin-wide water development with Pa Mong at its center—was a direct outcome of the technopolitical network that was set up through the Bureau's efforts, and it has persisted to present times. This section continues by tracing the construction, dynamics, and apparent unraveling of the technopolitical network focused on construction of the Pa Mong dam as a means of accomplishing geopolitical and developmental objectives in the Mekong basin.

The Genesis of an "American Project"

The US government's earliest engagements with development and the transfer of technical knowledge in the Mekong basin involved conception and coordination of a reconnaissance survey carried out by a team of Bureau experts in early 1956.[7] The State Department played a critical and largely hidden role in encouraging international cooperation in the Mekong region during this period.[8] Its officials steered the newly independent governments of Indochina toward cooperation with each other and, more importantly, with an American state willing to provide economic and technical assistance as part of an emerging strategy of containment in Cold War–era Southeast Asia. As early as 1954 US embassy personnel stationed in mainland Southeast Asia were urging Thai officials to consider regional economic cooperation under the rubric of a "Mekong River Authority."[9] John Mecklin, an official with the US Information Agency in Saigon, wrote a classified memorandum to Washington, DC, in January 1955 detailing what he called the "test on the Mekong," which constituted what he perceived as the geopolitical challenge for the United States in the region. Mecklin urged that the United States launch an "all out economic aid program" in Indochina to help fill the political vacuum left after the abrupt termination of nearly a century of French influence and to avoid Communist takeover of South Vietnam and thus "save" Cambodia and Laos as well.[10] The State Department subsequently requested from the Bureau a "team . . . experienced in river basin development" to undertake a reconnaissance mission in the Lower Mekong basin that should be "initiated as soon as possible."[11] US embassy officials in the riparian countries suggested that preliminary discussions of the survey should include "urgent implementation" of the study, and that if the study deemed the potential for water development sound, they should envision "hydro-electric development at some point on Mekong" along with comprehensive resource development in navigation, forestry, and agriculture.[12] After interagency hurdles related to expenses were overcome, and after discussions were held within the International Cooperation Administration (ICA) on how much priority should be given to Mekong development, three Bureau engineers were assigned to the survey team. These engineers represented several decades of experience across a diversity of water management and dam construction projects.[13]

A UN-sanctioned team composed of international experts completed an investigation in 1957, at about the same time as the Bureau study.[14] These missions and behind-the-scenes lobbying by officials affiliated with both the United States and the UN Economic Commission for Asia and the Far East (ECAFE)—an agency with a long-standing interest in river basin development—led to creation of the Committee for the Co-ordination of Investigations of the Lower Mekong Basin (Mekong Committee), following an agreement signed in 1957 by the governments of Thailand, Laos, Cambodia, and South Vietnam. Notably, the agreement did not include the upstream states of Burma (which had a negligible portion of the basin within its territory) and, more significantly, China, an omission that presents enormous hydropolitical challenges today. At a technological level, this agreement paved the way for the investigations that prioritized Pa Mong and additional large-scale projects, but the Bureau's early engagements in the region are also clearly indicative of the overarching role of geopolitical considerations in the unfolding Mekong-oriented technical and economic initiatives. An odd but revealing episode during the Bureau experts' initial research visit to the region illustrates this point.

One night in January 1956, Gerald Strauss, a mid-level bureaucrat working for the US Operations Mission (USOM) to Cambodia in Phnom Penh, was rudely awakened by his "servant" at 11 in the evening. William Kirby, the servant informed Mr. Strauss, was at the door and wanted to see him regarding an urgent matter. The matter under consideration, at first glance, seemed rather banal: Kirby, an adviser to the ICA in Thailand and neighboring countries, informed Strauss that the arrangements for the travel needs of the Mekong River reconnaissance team, consisting of the aforementioned Bureau engineers and assorted embassy personnel from the Mekong region countries (Laos, Cambodia, Vietnam, and Thailand) were inadequate. Kirby further informed Strauss that he had already chartered an additional plane to ferry two Cambodian delegation members and three United States Information Service (USIS) bureaucrats on a portion of the team's travels throughout the region. Strauss told Kirby that the US embassy in Cambodia was unauthorized to spend US tax dollars in this fashion, at which point Kirby "started getting rather unpleasant and indicated that this entire scheme had been worked out by the State Department and that one of the major purposes of the entire Reconnaissance was to provide for better harmony between the four countries involved." After some additional quarreling over the qualifications of the two Cambodians that Kirby wanted to join the junket, "Mr. Kirby," according to Strauss, "becoming more and more disagreeable, repeated the entire State Department policy harangue." Strauss concluded his memo thusly: "This is not the first time that Kirby has been a troublemaker in Cambodia."[15] Clearly, Kirby's "harangue" was a reminder to all concerned in the Mekong initiative not to forget the geopolitical roots of the ostensibly technical project, a priority that filtered down to all levels of the US foreign policy apparatus.

The information presented in the original three-month survey, released in 1956, was rather anticlimactic, with no mention of Pa Mong or other major hydroelectric schemes. One of its central conclusions was that the fundamental data for "orderly development of the basin" were largely non-existent, and that top priority should be accorded to basic data collection relating to "stream gaging, rainfall, mapping, hydrographic and topographic surveys, and soils."[16] Assessments of the basin's potential were cautious and modest; the report noted that a number of "small hydroelectric projects appear attractive for study" and that flood control, presumed by many basin development advocates to be a core goal of damming the river, "was of doubtful value" given that "most of the [riparian government] officials questioned stated that floods were beneficial to agriculture, fish production, and high water navigation."[17] Despite the Bureau experts' lukewarm appraisals of the immediate development potential and the lack of critical hydrological information, embassy officials and ICA staff members (à la William Kirby's diatribe) were already committed to promoting cooperative economic development of the Mekong as a key objective of US policy. As early as January 1955, ICA director John Hollister gave a speech in Bangkok in which he mentioned the "gigantic power [and] irrigation potential of Mekong benefitting four Southeast Asia countries" and touted the Mekong as an example of a project under study by the United States for long-term assistance.[18] Assessments of US strategic interests in mainland Southeast Asia also made the case for US involvement in Mekong development. A 1959 National Security Council report emphasized the "importance of steady economic growth and political and social stability in non-Communist Asia, if it is not to succumb to Communist pressures or lures," and noted bluntly that the "weakness and instability of various non-Communist Far East countries—including in some instances a lack of popular identification with the regime in power—engendered by the area's incomplete political, social and economic revolutions, are major handicaps in meeting the Communist threat."[19] This same report advised the US bilateral aid programs to support "regional and Free World cooperation advantageous to U.S. objectives through such measures

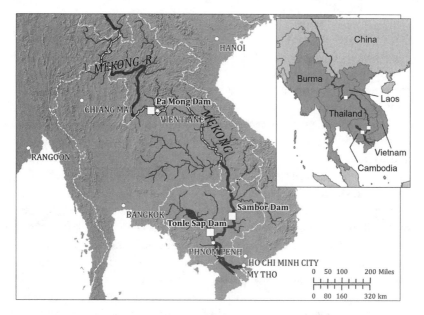

5.1. Mekong River basin development as conceived by 1956 Bureau report, which featured Pa Mong and two other potential dam sites (Sambor and Tonle Sap) in the basin. Source: US Bureau of Reclamation, *Pa Mong Phase II—Appendix 5, Plans and Estimates* (Denver: United States Bureau of Reclamation, 1972), 6.

as appropriate support of . . . regional undertakings such as the Mekong Valley Project" and to give special attention to "impact type projects."

Having established the Mekong River basin as the focus of US technical assistance in spite of the rather tepid assessment of the river's development potential by the Bureau's water experts, Mekong-oriented planners began the search for an "impact type" project to capture the imagination of the riparian states and attract the technical, financial, and geopolitical attentions of international donors, particularly the US government. The Pa Mong dam—to be located on the main channel of the Mekong River approximately 20 km upstream of the Lao capital Vientiane (fig. 5.1)—was first proposed by staff of the Mekong Secretariat, the administrative and technical arm of the Mekong Committee, as part of a comprehensive river basin development plan that anticipated eight massive hydroelectric dams on the river's main channel and a slate of tributary projects within the basin's stream network.[20] The Mekong Committee and its chief executive officer, C. Hart Schaaf, astutely perceived the attractiveness of a single massive project in terms of garnering US financial support and thus demonstrating US beneficence in the region. As signaled at the beginning of this chapter, the Pa

Mong project became solidified as the focal point of Mekong development, and of American attention, following the timely visit by Commissioner of Reclamation Floyd Dominy in 1961.

Dominy's personal outing to the region and the Pa Mong site further established the tight linkages between technological expertise and geopolitical aims in two important ways.[21] First, as noted above, he was practically ordered to Thailand by the State Department in order to demonstrate US commitments to the region's governments in the form of long-term economic and technical assistance as a counterpoint to growing US military involvement in Indochina. As recounted in a memorandum from the Bangkok embassy in July 1961, Dominy accomplished this in a masterful way by personally joining in an aerial reconnaissance of the dam site. He played up the Pa Mong scheme as a "tremendous project as large as if not larger than anything undertaken by the Bureau of Reclamation in the US," and he committed the Bureau to sending three experts to begin more in-depth investigations of the dam's feasibility.[22] Second, and echoing the sentiment directed toward John Savage nearly two decades earlier, Dominy clearly established the Bureau's credentials as the world's foremost water development agency, one willing and able to provide the appropriate technical expertise for what was one of the world's largest and most challenging projects. In the words of a US embassy official, the visit was highly successful, showcasing for Thai and Lao representatives "a highly qualified team headed by one obviously in full command of the science of putting water and land to use for economic development" and demonstrating "tangible evidence of US interest in the area."[23] Dominy wielded science as an instrument of political showmanship.

During this same visit, however, Dominy also advised the riparian governments of "the need to proceed carefully to determine the exact nature of the [Pa Mong] project," and (perhaps with lingering memories of the Bureau's concurrent Blue Nile studies) he "warned against overoptimism" regarding the speed at which development of the Mekong might commence. The reluctance of US technical experts to set a rapid timetable for completing Pa Mong and similar projects was not palatable to officials in the US State Department, who were convinced that technical assistance, if deployed rapidly, was a critical tool in the Cold War's ideological struggles. At roughly the same time as Dominy's Mekong excursion, Southeast Asian specialist Kenneth Landon advised the John F. Kennedy administration that the "Mekong has enormous potential for the political and economic future of Southeast Asia and great significance for the future of U.S. relations with the riparian countries." And yet efforts to "derive the maximum short-term political impact possible" from the administration's proposed funding of

Pa Mong studies were in danger of derailment. The source of this disruption was evident to Landon: A "major hazard lies within our own government [particularly the ICA] because of the one-dimensional thinking . . . *at the technical level* [emphasis added]." He bemoaned the fact that ICA staff, following conversations with Bureau experts, "concluded it would be premature from a technical standpoint to begin an engineering survey of the Pa Mong dam site" without due attention to economic analyses, alternative options for resource development, and the heretofore undocumented needs of the people in the region of the project. Ultimately, as Landon pointed out, the State Department perspective triumphed, and the Bureau studies were initiated under the authorization of US foreign policy officials. The geopolitical rationale for this triumph is telling: "Political considerations [are] overriding," and the offer to fund engineering feasibility studies "will have the value of tagging the Pa Mong site as an American project."[24]

In effect, the technical expertise, represented by the credentials and preliminary analyses of the Bureau's water experts, was proving intractable to the geopolitical calculations behind the rapid deployment of American aid. In response, those experts, who recognized the complexity of constructing even a single dam on a river whose dynamic characteristics had yet to be adequately measured, became further embedded within an emerging technopolitical network that prioritized geopolitical expediency. As a result, doubts over Pa Mong's technical feasibility withered away for a time in the middle 1960s. As the Bureau's studies of Pa Mong and its environs proceeded and the hydrological, geographical, and socioeconomic data supporting the case for the dam were collected, inscribed, and disseminated through Bureau channels, the technological hurdles receded further into the background. By the mid-1960s the technical work and water resource expertise devoted to the project had become so deeply entwined with the Pa Mong dam's geopolitical origins and ends that the technopolitical project gained momentum that was partially, if not wholly, divorced from its technological and economic feasibility.

Extending a Technopolitical Network: The "Sex Appeal" of Large Dams

Eventually the Bureau received approval, and the necessary funding, to carry out feasibility studies on the Pa Mong scheme, and in August 1963 it officially launched what would become its longest and most expensive overseas venture. By 1965 the Bureau had seven full-time engineering experts working on the Pa Mong studies, and specialists in soils, geology, drainage,

5.2. Bureau of Reclamation team (*left to right:* Harrison, Dalebrough, Binson, Hoffman, Tan, Bush, Wood, Jacobson, and two Lao guides) at Pa Mong dam site on Thai bank, November 3, 1961. Courtesy of National Archives, Denver, Colorado.

and economics were assigned for periods of varying lengths to what became known among its affiliated experts as the Mekong project. The Thai and Lao governments contributed over forty staff members.[25] Expenses related to the Pa Mong studies totaled roughly US$10 million over the approximately 12-year life of the program. Phase I, carried out from 1963 to 1966, generated an extraordinary amount of technical information pertaining to the feasibility of the massive dam from a biophysical and economic standpoint, although this seemed a foregone conclusion given the geopolitical value attached to the project. Phase II, initiated in 1966 and targeted for completion in an estimated five to seven years, was designed to produce a report on Pa Mong that would bring the study "to reconnaissance standards, defined as generalized estimates of costs and benefits of all aspects including irrigation, power production, flood control, and improvement of navigation from increased flows and reduction of salt water intrusion."[26] When fully completed, the dam's power plants were projected to produce between 4,800 and 5,400 megawatts (dwarfing regional energy demand at the time), and the dam's reservoir was expected to provide enough stored water to irrigate more than 1 million hectares in the central plains of Laos and in Northeast Thailand (fig. 5.2). According to a Bureau of Reclamation promotional brochure at the time, the reservoir produced by the dam would be "two and one-half times the size of the largest man-made lake in North America, improving navigation far upstream, as well as [providing] additional flows during low-water periods for navigation downstream of the project."[27]

It is impossible, conceptually and technologically, to separate construction of the Pa Mong dam from the broader program of water resource development envisioned for the Mekong's entire basin, a program that was supported by Mekong Committee staff, US foreign policy experts, Bureau engineers, and—at times haltingly—the basin's riparian governments. The general idea was to use Pa Mong and two additional massive dams (the Sambor and Stung Treng dams in Cambodia; see fig. 5.1) for the production of hydroelectricity to fuel industrialization in the region and, simultaneously, to use the enormous amount of water stored in their reservoirs to stimulate modern irrigation development. The Pa Mong scheme was "considered as the key one in the entire system,"

> not only because the Pa Mong project . . . would produce the largest block of power and . . . bring the largest area under irrigation in the semi-arid region of northeastern part of Thailand, but because of its vast storage capacity that can substantially increase the low water discharge six fold from about 500 cms to 3,000 cms [cubic meters per second]. Such an increase of perennial flow would add tremendously to the power production capacities of all four multi-purpose projects located downstream, as well as materially increase the [water flows] available for navigation and reduce the extent of salt water intrusion [in the delta].[28]

This perspective on Pa Mong as the catalyst for basin-wide water resource development was reiterated in a later interim report by the Bureau, which argued that Pa Mong must be considered "in the perspective of overall basin development . . . [that] applies not only to the integrated planning of Mekong projects but also to . . . development schemes such as transportation systems, industrial installations, and educational programs."[29] This viewpoint was captured well in an imaginative rendition of Pa Mong as the "sinew of development" that would bring electricity transmission, urban and industrial development, and modern agricultural systems to nearby landscapes (fig. 5.3).[30] Bureau engineers also had to undertake the delicate political task of negotiating, albeit within technical discourses, the complicated dynamics of the riparian states' differential interests in Pa Mong and their broader development goals in the basin. As early as 1961 the Thai prime minister expressed the hope that a high dam at Pa Mong would contribute to modern irrigation development in the country's northeast, which was habitually perceived as Thailand's most "backward" region and thus susceptible to communist

5.3. Pa Mong as the "sinew of development" as presented in a 1968 promotional brochure from the US Agency for International Development. Source: US Agency for International Development, *To Tame a River* (Washington, DC: USAID, 1968), 13.

influence. In a similar vein, Oukeo Souvannavong, the Lao commissioner of planning at that time—whose government was confronting growing insurgent forces and a host of internal political conflicts[31]—made the point to the Bureau that "it was extremely important for the dam to be high enough to provide gravity irrigation to Vientiane plain in Laos [sic]," and to observers of the 1961 meeting between Dominy and Oukeo, it was obvious that the Lao official "was trying to assess the seriousness of the American intention."[32]

The most publicly visible expression of the role of Mekong development in Cold War geopolitics was arguably its designation by the administration of Lyndon Johnson as an "alternative strategy" for bringing about peace in Southeast Asia during the latter half of the 1960s. Johnson and his advisers hoped that development of the basin—guided by Bureau experts and funded through USAID—would become a political and cultural symbol of peace and of cooperative efforts toward material progress for a strife-ridden region. He proposed a total aid package of US$89 million to help in the "peaceful development of Southeast Asia,"[33] declaring that the "vast Mekong River can provide food and water and power on a scale to dwarf even our own TVA."[34] Yet this aid package largely failed to materialize, in part due to growing concerns over the efficacy of large-scale river basin development in the Mekong basin and, more obviously, due to domestic objections in the US Congress and among the American public to continuing military involvement in Southeast Asia. For these and other reasons, Mekong development centered on construction of the Pa Mong dam became increasingly difficult to maintain as a viable exercise in technopolitics.

Nevertheless, Bureau staff continued to produce water-related knowledge and promote Mekong development as vital to US national interests. As noted rather grandiloquently by an American journalist working in the region, the "Bureau of Reclamation personnel, who have staked this out as their special area of action in the whole region, regard Pa Mong as a once-in-a-generation proposition whose scientific sex appeal would be felt by any virile young American engineer."[35] The engineers and associated planners working on Pa Mong and other projects seemed to assume that the scheme's efficacy was self-evident:

> Bedazzlement with the bright promise of Pa Mong . . . and other Mekong projects not infrequently infuses developmental enthusiasts with a sort of "project mystique" which induces them to accept certain ritualistic elements as though they constituted a set of positive values of their own. This means . . . the conversion of a suggested then often reiterated target date into a fixed schedule and the gradual acceptance of conjectures about costs and returns, once they

have been much mooted about, as firm commitments that the money must be raised and the results produced—the disposition . . . to make the transition, imperceptibly and inadvertently, from the contingent to the categorical. This may well be happening, for instance, with regard to Pa Mong.[36]

This move from the "contingent to the categorical" involved survey missions, a series of professional projections to US foreign policy officials and international aid experts, and most importantly, thorough technical analyses. The scale and scope of the Bureau's technical investigations undertaken as part of the Pa Mong program are incredible. The results of the phase II feasibility studies (at least their first stage) encompassed a series of massive volumes, covering a range of issues relating to the project's suitability—land resources; drainage; hydrology and climatology; geology—and including plans and estimates (detailing the figures for specific projects within the Pa Mong planning context, including a series of smaller dams and weirs to redirect flows for additional irrigation coverage); economic, agricultural, social, and financial analyses; and a series of corollary studies that dealt with, for example, fisheries and wildlife concerns. Throughout the different phases of the investigation, the Bureau's map (see fig. 5.1) of the dammed Mekong basin highlighting the most critical projects was reproduced hundreds of times in a multitude of reports (and became *the* basic representation of the "developed" Mekong well into the 1990s). One of the most significant aspects of such inscriptions is the "unique advantage they give in the rhetorical or polemical situation."[37] Thus the volumes and maps became crucial tools for demonstrating the Bureau's technical convictions regarding the feasibility and developmental benefits of the proposed projects. Additionally, as the multitude of graphs, detailed illustrations of the dam and dam site, tables, and other technical representation accumulated, resistance to the construction of Pa Mong on technical grounds, as recorded in earlier statements, became less salient.

However, as US officials began to comprehend the full costs of Pa Mong development, the project—which once appeared inevitable in the minds of nearly all the geopolitical and technical agents involved in its development— began to unravel, as did the technopolitical network that supported it. In 1968 project experts put the tab for the dam itself at US$600 million, although that figure did not include projected costs for irrigation infrastructure (potentially another US$1 billion) or indirect costs associated with training, maintenance, administration, education, and the like. Later estimates of full expenses, including transmission lines for delivering the dam's electricity and an array of irrigation projects, came to US$1.2 billion. While construction of the main dam and power plants was optimistically expected to be completed in five to ten years, building the irrigation component was expected to take twenty-five years.[38] By late 1967, in addition to growing concerns about the dam's economic viability and geopolitical utility (described below), Bureau staff members were expressing apprehension over whether geological conditions in the Pa Mong region were conducive to project goals for irrigation development, echoing earlier reconnaissance studies that identified problems with soils and land classification.[39] These and other issues led the Bureau, with the encouragement of State Department officials, to send senior staff to the Mekong region for "top-level guidance and direction [of] Pa Mong investigations" in an effort to see if such reservations could be addressed,[40] or if US sponsorship for the "big show" and the quintessentially "American project" should be curtailed.

Pa Mong in the 1970s: The Unraveling of a Technopolitical Network

It was becoming clearer by the late 1960s—while the Bureau's phase II operations continued to churn out biophysical and socioeconomic data on the project's location—that the relationship among Mekong development, the feasibility of the Pa Mong dam, the political dynamics within the region, and the geopolitical objectives of the US government had significantly changed. A Far East specialist within USAID observed in 1965 that although the United States had "stimulated the start of planning [in 1956] for a Mekong Development as a symbol and focal point for regional cooperation," the construction of Pa Mong and similar projects "would not be economically justified during the next decade." Moreover, the political dynamics within the region were changing in a way that was not conducive to the grand visions of large-scale development promised within Pa Mong planning documents. Political conditions and armed conflicts within Laos "made all project proposals in that country appear infeasible at present." Elsewhere, the "Cambodian government's expulsion of U.S. aid missions" made water resource development progress unmanageable, and the Vietnam delta was "a war zone," making projects there extremely unlikely. Given these conditions, the USAID synopsis argued that large-scale basin development over the next decade was "likely to be less significant to the economic progress of Thailand, Laos, Vietnam and Cambodia than many other immediately promising investments in their human, agricultural and industrial resources."[41]

By the early 1970s a host of US government officials—including President Richard M. Nixon—were refocusing foreign policy in Southeast Asia on how to effectively withdraw from engagement in Vietnam. In general, the Nixon administration placed far less priority than previous administrations had on the use of economic and technical assistance to hinder communist influence in the Third World.[42] In the context of Mekong development, this shift translated into a less direct role for US agencies, prominently the Bureau of Reclamation, and a greater role for multilateral entities such as the World Bank, the Asian Development Bank, and the various United Nations agencies that had historically been engaged in Mekong activities, such as the United Nations Development Programme. The developmental objectives for the Mekong basin also shifted: the focus on megaprojects, typified by the Pa Mong scheme, was partially set aside, in no small part due to the ongoing armed conflict and animosities among the riparian states. Instead, donor countries and the Mekong Secretariat temporarily placed the more ambitious mainstream development schemes in the background and directed their attention to development programs (e.g., agricultural training centers) that would ostensibly meet the immediate needs of the basin's largely rural populace. State Department and USAID communications began to discuss how Mekong development, including Pa Mong, might provide the impetus for "peacetime reconstruction and development of the Indo-China countries."[43]

The World Bank responded positively to a request by UN Secretary-General U Thant in mid-1969 that it "become more closely associated with the whole effort for the development of the Mekong basin in the new and critical phase of the cooperative effort now being entered."[44] The bank set up a Mekong Division in the final months of the year to review the numerous studies of Mekong development and develop future plans. At this point, a total of roughly US$70 million, provided by over thirty donor countries and an assortment of UN agencies, had been expended on investigating Mekong development. The bank's review revealed several problems, including continuing gaps in the data on basic biophysical processes in the basin, the varying quality of the numerous studies on Mekong development, a lack of coherence in terms of integrating the individual studies carried out by a panoply of different agents (i.e., many studies failed to build on or ignored previous work on similar topics), and a lack of coordination between the national development priorities of the riparian states and basin programs. This last problem was significant, as it had led to some unrealistic expectations on the part of the riparian governments regarding what kinds of development projects were feasible and could be implemented within a relatively short time. The bank's Mekong Division concluded "the time has now come to relate much more closely the diverse and disparate investigations, and to direct available resources into filling the more important gaps in evolving a realistic basin development program."[45]

The bank also proposed the establishment of a "common multilateral fund" to finance additional investigations. This fund would be managed by the World Bank to coordinate the expectations and responsibilities of the Mekong Committee agencies, the riparian states, and the donors. The fund, it was proposed, could be used for various types of investigations: general programming studies focused on how basin development would link to other parts of the riparian states' economies; major project feasibility studies such as the Pa Mong investigations; and "studies of specific areas or technical subjects of primarily local or specialist interest." Studies of major projects such as Pa Mong were to be de-emphasized, considering "prevailing conditions" in the region. Instead, the focus of Mekong development should turn toward tributary projects—which presumably require less attention to international concerns—and so-called pilot projects. The latter would focus on agricultural development, which was "generally regarded as the key to development of the area." Projects in agricultural development and the expansion of irrigation systems would "serve to provide relevant experience as a prelude to major schemes coming into operation later."[46]

As US engagement in Mekong development began to wane in the late 1960s, the basin's riparian governments—mirroring the perspectives of the Lebanese and Ethiopian governments in the face of stalled dam projects—expressed their growing dissatisfaction over the pace of promised projects. At a plenary session of the UN Economic Commission for Asia and the Far East (ECAFE) held in Phnom Penh in late 1969, "the member for Thailand took the Advisory Board and IBRD to task for suggesting a later date than 1980 for the completion of Pa Mong," although the Thai statement at the subsequent ECAFE plenary was "more subdued (and realistic)." At this same meeting, the Cambodian representative strongly endorsed the Stung Treng project, which represented a shift in Cambodia's position, as it had previously been inattentive to the massive scheme. One US embassy official in Bangkok saw this as significant given "the fact that Stung Treng may be one of the most important and attractive projects in the Basin from the standpoint of storage, power and regulation of the river."[47] Not surprisingly, the Thai, Cambodian, and Lao officials and development specialists that had invested an enormous amount of time and resources (from their perspective) in seeing the realization of Mekong dams were increasingly frustrated by the mismatch between the enormous volume of technical studies and the decided paucity of economic and political support.

By this time the technopolitical network constructed so carefully through the Bureau's water expertise and the State Department's geopolitical and developmental discourses was clearly coming apart. In addition to what might be best described as concerns *internal* to the Mekong development apparatus, more public discourses were expressing growing skepticism about the supposed benefits of large dams as well as awareness of the significant socioecological impacts of water projects such as Pa Mong. By the late 1960s a small but growing number of ecologists, anthropologists, and other researchers were generating critical analyses of the consequences of large dams and other "careless" technologies, particularly their effects on ecosystems and on resource-dependent livelihood patterns.[48] Representations of dams as a destructive technology, which had previously been rare, increasingly emerged within popular media outlets, and there was increasing pressure on the Bureau within the United States from environmental groups concerned over its lack of accountability for large water projects and their profound ecological and hydrological effects.[49] Claire Sterling, writing in the *Washington Post* about Pa Mong and similar projects, captures some of this awareness:

> When the bureau began its feasibility study [1962], billion dollar dams like Pa Mong were a straightforward proposition: if the dam could stand up, store water, and produce plenty of cheap power, why not? [And] those were the days when there were no ecologists underfoot and almost everybody thought big dams meant instant progress. Times have certainly changed.[50]

Sterling identified several of the costs of the US$1.1 billion Pa Mong project that had received little attention. The drowning of 948,000 acres of productive farmland, resulting in somewhere between US$57 and US$110 million in lost revenue, would be one of the more substantial costs associated with Pa Mong. Ensuring that the people displaced by the dam—numbering somewhere between 312,000 and 500,000, depending on completion of the project—would move to equally productive lands would involve prohibitive costs.[51] These and other factors lessened the viability of projects like Pa Mong and brought a new element—socioecological critiques and the social actors that served as their authors—into the diversity of networks assembled through large dams that has crucial resonance today (see chap. 6).

In what amounted to a last-ditch effort to maintain the technical, economic, and geopolitical forces that had previously driven Pa Mong and Mekong development forward, the World Bank and, more enthusiastically, the United Nations attempted to salvage the Mekong project. It was increasingly apparent in the early 1970s that the World Bank was expected to take a lead role in "reconstructing Indochina" following cessation of armed conflict in the region.[52] Furthermore, the United Nations held out hope for continued US involvement in Mekong development activities in the postwar period as US troops were withdrawn. In a 1972 conversation, C. V. Narasimhan (a chief adviser to the secretary-general and former executive secretary of ECAFE) expressed his hope to two officials from the State Department's Bureau of East Asian and Pacific Affairs that "development of the Mekong Basin could play a key role in the peacetime reconstruction and development of the Indo-China countries," and wondered about the role of the United States in "post-war reconstruction and development" in the region. Marshall Green responded that the United States "was deeply concerned about this subject" and indicated that President Nixon "might be prepared to contribute funds up to $7.5 billion, subject to Congressional approval, of course, for reconstruction once a satisfactory negotiated settlement was achieved."[53]

And what of the Bureau and its stake in Pa Mong? In 1969 Assistant Commissioner Gilbert Stamm delivered a speech to the Southeast Asia Development Advisory Group (SEADAG), a group of American scholars and professionals interested in Southeast Asia, on the tremendous benefits of Pa Mong and similar projects. After detailing the tremendous capacity of Pa Mong to improve lives with its hydropower generation, water storage and irrigation capabilities, flood control, and ancillary benefits, he dwelt on the international institutional arrangements among the four riparian countries of the Lower Mekong basin that would be necessary if the basin's water resources were to be equitably shared. He concluded the talk by underscoring that Pa Mong would undoubtedly "produce broad benefits for the people in the Mekong River's area of influence." In a revealing addendum to the official script of his talk, Stamm also archived his notes from the SEADAG meeting, noting at one point that most of the Pa Mong discussion could be boiled down into a "package" of institutional and geopolitical measures that would guarantee moving forward with Mekong development. Ultimately, this package needed to contain enough financial and political incentives that it could be "sold to bankers" and "accepted by riparian states." As the deliberations within the World Bank make clear, the marketing strategy failed, the hoped-for financial assistance never arrived, and the Pa Mong Dam effectively died.

Conclusion

The history of the Pa Mong dam serves as an exemplar of the critical amalgamation of technical expertise and (geo)political practices that animates so many large-scale water development projects. Geopolitical practices, in the first instance, allowed for the diffusion and application of technical knowledge and, in the second, actively shaped the intensity and duration of knowledge production. In terms of geopolitics, the *idea* of Pa Mong became more important, and eventually more powerful, than the millions of dollars expended on the studies and the Bureau expertise required to produce them. The architects of US foreign policy had, from the late 1950s onward, encouraged the rapid implementation of the Pa Mong project and development of the Mekong basin for a geopolitical end: a more stable Southeast Asia that gazed toward the United States rather than the Soviets as a model of political and economic modernization. The Bureau's ability to serve as a wellspring of technological know-how and, in Floyd Dominy's words, to present a "show" of technical expertise was absolutely critical for this geopolitical agenda. The Bureau's presence helped solidify political support for US interventions among the Mekong basin's riparian states. Yet, ironically, it was the relatively rapid shift in geopolitical conditions, highlighted by increasing American distaste for continued strategic engagement in the Mekong region, that ultimately overruled technical assessments of the dam's feasibility and effectively shelved the project.

But what else happened when the plans for Pa Mong unraveled in the face of seemingly insurmountable economic and geopolitical hurdles and (increasingly) critiques of its likely environmental and social costs? My response serves as a crucial argument of this book: the technical and geopolitical networks that Pa Mong played such a crucial role in assembling have morphed over the decades, but remain largely *in place* in the present day. The focus of these networks may be directed toward different projects, but the ideologies of development and the technical plans for an altered Mekong have proved remarkably resilient. Despite the power of geopolitics to determine past and present efforts to transform the Mekong into a tightly controlled and managed basin, those efforts all rest on a technical knowledge base first mobilized under the auspices of the Bureau of Reclamation and its colleagues during the late 1950s and 1960s. The Bureau's Pa Mong investigations—backed by its global reputation as the world's premier dam-building organization—generated enormous quantities of data and multiple inscriptions, nearly all designed to facilitate and represent the transformation of the Mekong into a "working" river by means of human manipulation of flowing water. Pa Mong was thus decisive for constructing an image of the Mekong as a river basin, as *water*, ripe for development and improvement if only the appropriate technologies and knowledge could be applied—an imagined geography that, while sustained by different actors, has persisted until more recent times and certainly drives current development initiatives (see chap. 6).[54] While extensive water resource development of Mekong tributaries—especially in Northeast Thailand and to some extent in the highlands of Laos and Vietnam—proceeded throughout the 1960s and 1970s, the Bureau's expert advice regarding Pa Mong and its prodigious hydroelectric capacity finds its current legacy in revived plans to build a series of large schemes on the river's main channel. Crucially, Floyd Dominy's "big show" of Pa Mong's transformative capacities and the constant refrain from US foreign policy officials throughout the 1960s regarding the project's efficacy inflated the developmental expectations of the riparian governments, which in turn fuel present-day desires to transform the Mekong. Moreover, the technical expertise originating in Bureau studies was intimately coupled to the geopolitical and economic alliances within the riparian states of the Lower Mekong basin. Even though these alliances were to a great extent manufactured by US officials and their counterparts within the United Nations and international financial institutions, and although they have no doubt fragmented and shifted over the decades, the shared vision of a dammed Mekong has remained robust. The hydroelectric schemes now planned for the main stream rely on the same imagined geography of the Mekong basin—a developmental vision of water control, symbolized as "progress" and "modernization"—that Pa Mong helped produce and which has proved remarkably resilient in the face of the complex geopolitical changes in the Mekong region.

The assemblage of networks represented by Pa Mong is also interwoven with a particular scalar politics. Discussions surrounding Mekong development during the 1960s were predicated on the understanding that Pa Mong, under the guidance of the world's most knowledgeable and effective water resource development agency, would usher in an era of basin-wide development and hence economic modernization. This was a scale-making project of a categorically different scope than previous Bureau efforts. In tandem with other large mainstream schemes, and as noted above, Pa Mong established and enhanced an imagined geography of the Mekong basin as a completely regulated river system. Yet this imagined geography was extraordinarily fragmented in the way it conceived the basin as both a biophysical and geopolitical entity. In the first instance, the project virtually ignored the basin's complex biophysical dynamics (including such matters as its upstream-downstream relationships, annual flooding cycles, and tremendous fisheries productivity) and the livelihoods of basin residents who depended on these dynamics.[55] Moreover, the basin, as apprehended within the contours of the Mekong project of the 1960s, simply cleaved off the upper basin situated within Chinese territory, effectively leaving the difficult international negotiations over how to share the *entire* basin's water resources to later decades and different governments. As the next chapter shows, however, the imagined geography of the Mekong basin has also shifted over the years to include consideration of its biophysical integrity *and* geopolitical boundaries as a whole spatial entity. Moreover, the power of the technopolitical network that has historically stimulated Mekong development has been challenged by new arrangements of human and nonhuman actors that make it possible to think of assemblages of water-society relations in ways that do not place large dams at their center.

SIX

Large Dams and the Contemporary Geopolitics of Development

Every historical era is . . . multitemporal, simultaneously drawing from the obsolete, the contemporary, and the futuristic. An object, a circumstance, is thus polychromic . . . and reveals a time that is gathered together, with multiple pleats.[1]

—Michel Serres

One of this book's central arguments is that large dams generate and hold together assemblages of geopolitical and technological networks—among others—that *linger*. The technopolitical networks catalyzed by the Bureau's work in China in the 1940s, in Lebanon in the 1950s, in Ethiopia and the Mekong basin in the 1960s, and its myriad interventions throughout the tricontinental world during the twentieth century's concrete revolution created a model of action and a pattern of thought that, fueled by the Bureau's technical expertise, played a crucial part in the damming of the world's rivers. Moreover, these networks persisted *regardless of whether the dams that resided at their core came into existence or not.* As noted in the previous chapter, this was most obvious in the case of the Pa Mong dam, which never materialized. It instead bequeathed a legacy of technical knowledge (in the form of reports, designs, and a multitude of social and biophysical calculations) and, more importantly, a hegemonic idea of the "developed" basin that has found its realization in present-day plans to dam the Mekong's main channel. Large dams, perhaps more so than any other so-called development intervention, transform socioecological systems in ways that not only create novel geographies of development (e.g., alterations in the spatial distribution of water, pathways of electricity generation, basin-oriented water governance), but also produce novel temporalities of development. These temporalities, which are extremely important in interpreting the contemporary geopolitical landscape of large dams and water development, encompass at least three dimensions.

First, the technopolitical networks mobilized by dams, as in the cases of the Pa Mong dam (chap. 5) and, to some extent, the Yangtze Gorge project envisioned by John Savage (chap. 2), endured as resources for governments, development planners, water experts, and the global dam industry to draw on far into the future. The technological scaffolding of these networks, once laid down in the form of feasibility reports, technical drawings, economic justifications, and so on, even in cases in which the projects and river basin schemes were not implemented, held a latent power. The power of the technical knowledge could thus be "switched on"—relatively quickly in some cases, such as the Karoun and Finchaa projects—wherever the political will of government agents interested in modernizing their river basins could be found. This will has hardly been in short supply, in decades past or present, although there are now powerful counter-narratives to dam-based water development from both social and ecological quarters.

Second, dams themselves persevere as material objects with profound socioecological consequences over time scales that generally extend well beyond the human agency, political-economic motivations, and technical expertise most directly responsible for their construction. As a result, they continue to act as obstructions to flowing water and contribute centrally to human modification of landscapes at multiple scales. However, dams do not "live" forever, and recent debates over dam removal have brought into sharp relief the question of the life span of dams (explored in chap. 7).

Finally, in a genealogical sense, the (geo)political, economic, cultural, and technological conditions that produce large dams are likely to change over their long life spans. Yet the *idea* of dams—as engines of growth and as means of controlling unruly rivers—has remained remarkably resilient in the face of these drastic changes. Moreover, as broader societal conditions have shifted over the decades, new arguments have been added to the litany of rationalizations for large dams and water resource development in general.[2] A prime example is the discourse of renewable energy and the now often-heard contention that hydropower is a critical element of the global effort required by states and societies to stave off and mitigate against the effects of a changing climate.

These observations on the temporality of large dams and, more broadly, river basin development provide an advantageous window into the contemporary geopolitics of large dams.[3] This chapter's central premise is that despite the presence of a surprisingly effective and politically shrewd global anti-dam movement, and despite some radical changes in the central human agents involved in promoting and disseminating large dams, the geopolitical conditions of the present day remain remarkably similar to those of the middle of the twentieth century in terms of their capacity to provide the ideological and financial wherewithal to continue damming the world's rivers. In other words, the technopolitical networks brought into being and stabilized around large dams remain remarkably potent agents of river alteration. And this premise helps put into a different light the obvious question of whether or not we have learned anything from the history of large dams and river basin development that might be fruitfully applied to contemporary debates over energy production, climate change, and nature-society relations. But who the "we" is in that simple statement matters critically. Of course the critics of large dams emanating from academic and advocacy circles, with support from the findings of the World Commission on Dams (see chap. 1), can point to any number of large-dam projects that have wreaked social and ecological havoc on river systems and riverine communities. One need only look to recent explications of the environmental and social history of projects like the Cahora Bassa Dam on the lower Zambezi River in Mozambique to find large dams that are indivisible from networks of colonialism, brutal labor exploitation, geopolitical intrigue, economic nonperformance, and violence against disenfranchised peasants and ecosystems.[4] Yet the Mozambican government, far from expressing wariness over the legacies of Cahora Bassa, is currently seeking financing for an even larger project (the Mphanda Nkuwa Dam) to be constructed on the Zambezi just a scant 60 kilometers downstream from Cahora Bassa. For the governments, corporations, and international financial institutions that are the main proponents of dams and hydropower, the technopolitical conditions of the present era are as generative as ever for an acceleration of building dams and transforming river basins. In effect, targeting large dams as the root of hydrological evil neglects these fertile technopolitical conditions because such a strategy—while certainly indispensable and in many instances effective—ignores, almost by political necessity, the broader and very powerful networks that large dams assemble.

I proceed with, first, some short examples that explore echoes of earlier chapters. The specters of the development of the Litani and Blue Nile basins, and perhaps most conspicuously the Mekong, continue to haunt current discussions regarding the geopolitics of large dams.[5] In these three cases the specters—identified with Serres's "pleats" within a polychromic universe— are particularly lively. Present-day programs of water resource development in these basins underscore how shifting global geopolitical alignments and environmental concerns over climate change and renewable energy have contributed to a new (global) era of dam building. We are, I argue, on the cusp of a second phase of the concrete revolution. Next I use these cases to explore the contemporary geopolitical architecture of dams and development in more detail, paying particular attention to the positions that the Chinese government and specific Chinese agencies have assumed within a global technopolitical network of dam financing and construction. To a large degree, Chinese agencies have assumed and expanded the technical and geopolitical roles that the Bureau of Reclamation played in the middle of the twentieth century, and I argue that their involvement has important repercussions. Picking up on themes highlighted in the work of the World Commission on Dams (WCD) and the decades-long struggles of the global anti-dam movement, I end the chapter by examining the political ecology of large dams and river alteration, emphasizing how the global anti-dam movement, debates over "clean" hydropower and climate change, and scientific understandings of river systems help define the current technopolitics of large dams. This discussion sets the stage for a conclusion (chap. 7) that explores the ways in which scholars, practitioners, and communities concerned about the likely socioecological effects of a renewed emphasis on large hydroelectric dams might construct alternative technopolitical networks focused on water and livelihoods as opposed to dams and capital.

For several reasons, this chapter is more conjectural than previous ones. There is no specific storehouse of records from which to draw the experiences of engineers or the strategizing of foreign policy advisers of the present era. The contemporary landscape of hydropower development is exceedingly dynamic, involving thousands of actors operating at global to local scales. The Chinese government agencies and private firms that are spearheading a new era of dam building are less accessible than a full-fledged empirical study might hope for. Still, my aim is to provide a glimpse of the geopolitics of large dams in the present day. In essence, it is a genealogy of the present, probing the geopolitical and political-economic conditions that encourage the global proliferation of large dams and river basin development in the face of mounting criticism over this enduring technological intervention.

"The More Things Change . . ."

In March 2011 the Ethiopian government announced plans to move forward with the Grand Ethiopian Renaissance Dam (GERD), a massive hydroelectric dam located on the Blue Nile (Abbay) River near the Ethiopia-Sudan border, designed to generate over 5,000 megawatts of electricity and store over 60 billion cubic meters of water in its reservoir.[6] To pay for the estimated US$4.8 billion scheme, the government announced later in 2011 that it would sell "GERD bonds" both domestically and abroad, in part to draw on the "patriotic sentiment" of the Ethiopian diaspora, who were expected to applaud this effort to develop their homeland.[7] Not surprisingly, the Egyptian government criticized the scheme because of its likely effects on water flows in the downstream reaches of the Nile basin and the Ethiopian government's seeming lack of consultation with its riparian neighbors. By 2013, as the dam's reservoir started to fill and Egypt again raised concerns about the "cutting off" of the Nile's waters, an array of commentators and analysts invoked the rhetoric of "water wars" to signal this seemingly new era in the hydropolitics of the Nile River basin. Ethiopian officials shrugged off Egyptian criticism of the project and concerns that the Nile's downstream flows would be reduced by roughly 15 percent during the filling of the reservoir; one civil engineer enthusiastically compared the GERD to Hoover Dam, hoping that "Ethiopia's dam can achieve for his country what Hoover Dam did for the US."[8] In a cursory nod to history, one expert noted that the dam itself "has been on the drawing board since the 1960s."[9] Unsurprisingly, the site of the GERD project was originally identified as a site for future ("next century") hydropower development in the Bureau's 1964 report, when the project was known as the more nondescript Border Dam. Many of the concerns over the regional hydropolitics of the Nile—acknowledged by State Department officials and negotiated by Bureau experts in the 1950s—are coming to fruition, albeit not precisely in the way originally imagined (see chap. 4).

The Ethiopian government has focused development attention on other parts of the Blue Nile drainage area as well. With technical and financial support from China, the government completed the 460-megawatt Tana Beles hydropower scheme (located adjacent to an outlet from Lake Tana and emptying into the Beles River) in May 2010, and it had previously completed another major hydropower dam on the Tekeze River in the basin, with central involvement from two prominent Chinese water development firms.[10] In another direct link to the Bureau's reconnaissance studies, the Ethiopian state has plans to move forward on the Kara Dobi hydroelectric dam, at a location that was pegged in the 1964 report as a "promising" site for hydroelectric development. Finally, Chinese actors have been centrally involved in providing technical support and financing for the Finchaa-Amerti-Neshe (FAN) scheme, an expansion of the one project actually constructed as part of the Bureau's work in the Blue Nile basin. This upgraded scheme, involving additional dam construction on two tributaries of the Finchaa River, was financed through China's Export-Import Bank and built by the China Gezhouba Group Company, one of China's (and hence the world's) leading engineering companies, whose record includes a prominent role in construction of the Three Gorges Dam.[11]

The Middle East, and specifically Lebanon, is also becoming embroiled within the latest global push toward hydropower and water resource development under the auspices of large dams. Confronting recurring water shortages, especially in its arid Bekaa Valley region (the location of the Karoun Dam), the Lebanese government has for several years tried to overcome financial and political hurdles in order to expand its water infrastructure. In December 2011 the Iranian government agreed to provide funding for the US$40 million Balaa Dam (currently under construction) located near the city of Tannourine in the mountainous region north of Beirut. There are two dimensions to this announcement that connect the Lebanese government's water development plans to both past and present technical and geopolitical networks. First, this project—along with active and proposed dams throughout Central Asia (Tajikistan, Azerbaijan, Armenia, and Kyrgyzstan) and Latin America (Nicaragua and Ecuador)—is seen by some observers as part of the Iranian government's "dam diplomacy," an effort to use technical and economic assistance for the development of water resources as an instrument to curry international and regional favor within a hostile global geopolitical environment. For the Balaa project, Iran originally stipulated that an Iranian firm be awarded the construction contract. The Iranian government dropped this condition after Lebanese concerns arose due to the close linkages between Iranian construction firms and Khatam al-Anbia, the economic wing of the politically influential Islamic Revolutionary Guard Corps.[12] The site of the Balaa project is within one of the roughly dozen river basins draining into the Mediterranean Sea that were sites of Bureau of Reclamation reconnaissance studies—in addition to its primary work in the Litani—in the mid-1950s.[13]

Second, as in the case of the Litani basin scheme and Karoun Dam decades earlier, internal political struggles have stymied Lebanon's efforts to greatly expand development of its water resources. A journalist reported in early 2013 that a combination of poor planning, lack of resources and political will, and emerging concerns over social and environmental disruptions have prevented the government from addressing what experts predict will be a serious water shortage within the country.[14] The Ministry of Energy and Water, charged with developing the nation's water resources, is chronically understaffed and underfunded, and its relationship with the Litani River Authority (LRA) is unclear. The LRA, created as an institutional doppelganger of the Bureau of Reclamation during the Bureau's Litani activities in the 1950s, has ambitious plans to extend hydropower operations throughout the basin and irrigation development to the Aita Shaab region in the south. Concerns linger among all levels of Lebanese society that ultimately the Israeli state—identified by some as a regional hydro-hegemon—will attempt a "water grab" for the Litani's southern reaches to address its own rising water demands.[15] Thus the existing technopolitical conditions for water resource development in Lebanon—revolving around regional geopolitics and internal institutional struggles—remain remarkably similar to those found by Robert Herdman and his Bureau team in the 1950s.

A cascade of mainstream dams being contemplated in the lower Mekong basin, combined with the existing hydroelectric dams on the main channel of the upper Mekong (known as the Lancang River) and the Chinese government's decision to build several more, will transform the basin's biophysical relations and result in an irrevocably altered geography.[16] Most recently controversy has erupted over the plans of the Lao government—which represents itself as the future "battery of Southeast Asia"[17]—to construct the massive Xayaburi Dam on a stretch of the Mekong within its borders. While the project's feasibility from an engineering point of view is unquestioned, critics assert that the Thai-financed dam—slated to produce nearly 1,300 megawatts of electricity, nearly all to be sold directly to Thai utilities—and the planned construction of eleven additional mainstream projects will almost certainly impair the basin's fisheries and the millions of rural livelihoods that depend partly or wholly on those fisheries.[18] Elsewhere in the basin, Chinese companies (e.g., Huaneng and Sino-Hydro) heavily involved in the current spurt of global dam construction are offering technical assistance and investment capital to ongoing tributary projects in the Lao PDR and Cambodia, and they remain key players in the proposed development of up to twelve mainstream dams on the Mekong.[19] In sum, Chinese engagements in the Mekong region signal a novel set of geopolitical and political-economic conditions for development of the basin. These circumstances are vastly different from the geopolitical and technical networks that generated the Pa Mong project and the early Mekong hydro-dreams, but the end result, if carried out, will be the same: a "tamed" Mekong subject to nearly complete human control.

Recent Chinese efforts to use economic and technical assistance in the Mekong region as a means of garnering geopolitical influence have prompted a plethora of evocative commentaries about the rise of China as a regional hegemon.[20] The US government, in a partial reverberation of its previous engagements in the region, has insinuated itself into current debates over Mekong development. In August 2010 Secretary of State Hilary Clinton, as part of a wide-ranging visit to Southeast Asia, met with the foreign ministers of the Lao PDR, Thailand, Vietnam, and Cambodia in Hanoi to endorse the activities carried out under the Lower Mekong Initiative (LMI), a program created in June 2009 as a way to enhance cooperation among the United States and the basin countries in the areas of health, education, infrastructure development, and environmental governance. As Secretary Clinton announced at the 2010 meeting, the "United States is back in Southeast Asia . . . and we are fully engaged with our . . . partners on the wide range of challenges confronting us."[21] This reengagement of the US State Department with basin governments to serve as a counterweight to Chinese power has been met with broad approval in some Washington foreign policy circles.[22] While technical support is a prominent part of the LMI, particularly in the arenas of health, education, and communications, no mention is made of dams or water resource development. One policy think tank representative, noting that "infrastructure remains a blank space" within the goals of the LMI, suggests that the United States should officially "oppose egregious hydropower projects [in the Lower Mekong], especially mainstream dams" on the river. This same analyst recognizes that the United States "should not, cannot, and does not seek to compete with China for infrastructure assistance," but must instead "use our expanded engagement with the region to 'keep China honest.'"[23]

In numerous river basins of the tricontinental world, governments are contemplating and carrying out programs of dam building at a startling scope and pace. Global headlines are replete with stories of the hundreds if not thousands of dam projects being planned to meet the increasing energy demands of the "developing" world, and environmental scientists are drawing attention to what they see as a distinct quickening of the pace at which states are transforming river systems in the name of economic development. The governments of India, Nepal, Pakistan, and Bhutan foresee the construction of more than 400 hydroelectric dams in their respective portions of the Himalayas (expected to produce a staggering 160,000 megawatts of electricity), and China is expected to build an additional 100 projects on rivers that flow out of the ecologically and politically significant Tibetan region.[24] Ecologists expect that this "water grab" in the Himalayas will cause massive social and ecological upheavals and exacerbate tensions over transboundary waters.[25] Alongside the Chinese government's decades-long program to exploit nearly every economically feasible hydroelectric site within its national territory, a more focused hydropower construction program in its southwestern provinces of Sichuan and Yunnan and parts of Tibet on multiple rivers—including those in the upper reaches of the Mekong and Salween basins—has provoked multiple instances of civic unrest in the largely rural communities bearing the brunt of hydroelectric development. What one commentator called the "most aggressive dam-building program in history" includes the construction of over a hundred large dams on tributaries of the upper Yangtze alone.[26] In South America, the governments of Peru, Ecuador, Bolivia, and Colombia are planning to build 151 large hydroelectric dams over the next twenty years in the Andean Amazon. This wide-scale alteration of rivers and affiliated infrastructure development (e.g., new road construction, electricity transmission lines) will lead to increased fragmentation of the region's river systems and have significant impacts on its biota.[27] These brief examples could be bolstered by enumeration of similar hydropower development programs being planned and executed throughout Africa, Southeast Asia, and the Middle East, offering ample evidence that the "concrete revolution" is far from over.

The renewed effort on the part of many of the world's states to intensify and hasten the transformation of river basins is enabled by an international political economy that provides more and more productive ground for the implementation of large-scale water infrastructure. Given the recent public discourse around hydropower emanating from international financial institutions, governments throughout the Global South, and representatives of large, infrastructure-centered engineering firms, it is hard to avoid the general impression that we are entering a new era of damming the planet's rivers. There are several critical signals that this is occurring, but perhaps none are more important than those emanating from the World Bank, the organization that since its inception in 1944 has arguably had a financial or planning role in hundreds if not thousands of large dams scattered across Asia, the Middle East, Africa, and Latin America.[28] Although the bank's investments in water-related development infrastructure—and indeed, public sources of investment capital in general—have been outstripped in recent decades by private sources,[29] World Bank financial engagement, at any level, in hydropower projects sends a clear signal to private investors that a project is economically "sound" and ostensibly meets the minimum acceptable levels of social and environmental impacts as defined in bank guidelines for hydropower development. In stark contrast to the cautious attitudes toward dams adopted throughout the bank's participation in the WCD process, a 2009 bank report titled *Directions in Hydropower* positively glows in its assessment of the role that hydropower development will play in "poverty alleviation and sustainable development" in the twenty-first century.[30] The report notes the "abundant physical and engineering hydropower potential in developing countries," 70 percent of which remains unexploited.[31] The report concludes that conditions in the developing world call for a dramatic "scaling up" of hydropower development through increased financing and capacity building for interested governments, while certainly keeping in mind the need for socially and ecologically sustainable forms of imple-mentation.[32] What is perhaps most extraordinary about this report, and a clear reflection of the importance of rhetoric, is the utter lack of reference to "dams" throughout the length of the document. Instead, the report high-lights "hydropower" and "multipurpose water infrastructure" as crucial an-tecedents of development in the tricontinental world. This wording fortifies the notion that large dams are better represented as *assemblages* of networks than as technological objects in and of themselves. In a sense, the World Bank, by refusing to name dams as dams, props up the specific technopo-litical network that international financial institutions assume large dams pull together. This network, in the bank's view, consists primarily of the electricity/hydropower produced by rivers (and their dams) and the other ("multiple") benefits of water storage and flood control (described below). Social and environmental concerns are of course mentioned in the report, but they are relegated to secondary status. It almost goes without saying that the geopolitical dynamics that continue to swirl around the conception and implementation of large dams are ignored.

More recently, and more frankly, a World Bank representative—at an in-ternational conference on hydropower development held in 2013—called for a new era of hydroelectric development spearheaded by the bank in line with its long history of engagement with large dams, although again repackaged as "hydropower." Rachel Kyte, the vice president for sustainable development and a prominent member of bank president Jim Yong Kim's advisory circle, argued forcefully that "large hydro is a very big part of the solution for Africa and South Asia and Southeast Asia. . . . I fundamentally believe we have to be involved." The "solution" Kyte refers to is the interna-tional effort to tackle the lack of affordable energy production options in the developing world. "Large hydro" is a relatively cheap and renewable energy source that, in addition to catalyzing industrialization, is "clean" in terms of greenhouse gas emissions (although this point is contentious, as we will see below). The bank's recent reticence over funding and supporting large-dam projects, evidenced by its involvement in the WCD in the late 1990s, was, according to Kyte, a mistake. Caution on hydropower development is "send-ing the wrong message. . . . That was then. This is now. We are back."[33] To the global hydropower industry, this was an encouraging message. A brief-ing published by the International Commission on Large Dams (ICOLD), a trade organization representing the world's largest dam construction firms and assorted government agencies, noted gleefully that at recent conferences on the future of hydropower, there was almost no mention of the WCD and its recommendations, even from representatives of the arch-nemesis of large dams, the Berkeley, California-based nongovernmental organization (NGO) International Rivers.[34] Clearly large dams—arguably never having receded as much as supporters of the WCD process might have wished—are a central part of the international development agenda once again.

These brief synopses of the contemporary (geo)politics of dam building highlight several critical trends in the early years of the twenty-first century. Not least, many of the most important actors and institutions that have the financial and political capacity—and a clear developmental directive—to move a global dam-building agenda forward are flexing their institutional, scientific, and rhetorical muscles in the name of large dams. Governments as diverse as those of the Lao PDR, Ethiopia, Pakistan, Lebanon, Brazil, and Cambodia, no doubt bolstered by the pronouncements of international financial institutions such as the World Bank and a renewed interest from the private sector in funding water infrastructure, perceive large hydroelectric dams as critical engines of industrialization and general economic development. Moreover, the idea that the construction of large dams is a vital and preferential way for nation-states to develop their water resources in the service of industrial, agricultural, and other water use demands remains a potent hegemonic concept within water governance institutions.[35] Along with marketization and scarcity discourses, the notion that the multiple social and ecological disruptions caused by large dams are necessary trade-offs against their benefits—"clean" and renewable energy, stored water for agricultural and urban uses, mitigation of floods, and enhanced navigability of rivers—is in many government and development bank quarters unassailable.[36] At the conclusion of the twentieth century, the WCD seemed to shift the debate over large dams momentarily to a question of whether or not dams actually fulfill the human development ideals they ostensibly promote (e.g., improved well-being, enhanced economic activities) and to hold up the possibility of more just and sustainable non-dam alternatives.[37] By contrast, the debate appears to once more reside within the field of influence of large-dam proponents, and the question is not "To dam or not to dam?" but rather "How can we mitigate the unfortunate but necessary impacts of large dams?" The continued damming of the planet's rivers appears a fait accompli.

Yet these vignettes also raise a question that is more directly resonant with the themes explored in this book: How can the Bureau's activities and their often opaque linkages to Cold War geopolitics possibly help us understand and think about the contemporary geopolitics of large dams and water resource development? The answer, not surprisingly, lies within the power of large dams to assemble and maintain diverse networks of nature-society relations. I argue that paying attention to the technopolitical networks first constructed throughout the 1950s and 1960s (see chaps. 2 through 6) directs critical attention to the contemporary discourses and practices of water resource governance and development and to the persistence of networks of deeply interwoven geopolitical dynamics and technical expertise. The novel, basin-oriented geographies envisioned and represented by Bureau experts working on the projects discussed in previous chapters were in some cases accomplished decades later (e.g., the Yangtze Gorge/Three Gorges) or partially fulfilled through the construction of multiple tributary projects within the national territories of the basins' riparian states (e.g., the Mekong basin). In other cases (e.g., the Litani and Blue Nile basins), the Bureau's geographical visions are in the process of being realized. The language of river basin development is not as prominent as in the prime of the Bureau's overseas activities, but the various scale-making projects and geographical imaginaries at work—ranging from *global* hydropower investment to *regional* geopolitical sensitivities—are as vigorous as ever.

The New "Dam Builders to the World"

The emergence of China as a global economic power over the past two decades has revolutionized the institutional landscape of development assistance and its potential geopolitical applications.[38] As a growing number of researchers and commentators have concluded, Chinese economic and technical assistance programs to the tricontinental world are now approaching or even surpassing the levels of aid offered by the traditional aid-donating nations of the Organisation for Economic Co-operation and Development (OECD), and they are reshaping the global architecture of North-South development relations.[39] Chinese aid is, for many recipient nations, of a qualitatively different nature: economic assistance is ostensibly more transparent, since any conditions attached to its disbursement are primarily economic and not directed at political and institutional improvements (e.g., "good governance") within the recipient nations. The Chinese state is not the only nontraditional aid donor, as the governments of India, Brazil, South Korea, South Africa, and many others associated with the Global South are using aid as a means to exert global and regional economic and political influence.[40] Africa has been a central target of aid (and trade) for China and other emerging donors, and Chinese aid flows directed toward the continent have increased exponentially in the past decade. Much of this aid has been targeted at development of mineral and energy resources within the recipient countries with a mind to securing strategically important resources.[41] Obviously there are enormously important questions emanating from the changing global landscape of foreign assistance and its potential to recalibrate power relationships along both North-South and South-South axes, but I would like to use these issues as an entry point into thinking about what this changing landscape might mean for the contemporary geopolitics of large dams. This intention implies closer scrutiny of Chinese economic and technical assistance as it pertains to large dams and hydropower development. With this goal in mind, I highlight two salient elements of Chinese companies' incipient role as the new "dam builders to the world"[42]: the geopolitical rationales propelling this trend (mindful of its political economic motivations as well) and, commensurate with the Bureau's history, the functions of technical assistance.

If the Bureau of Reclamation served a key role in the global dissemination of technologies and ideologies associated with large dams and river basins during the twentieth century, that role has been almost entirely supplanted by Chinese parastatal companies such as Sinohydro Corporation and other dam-building agencies. As of August 2012 Chinese companies or financiers were engaged with at least 308 dam projects, nearly all for hydropower, in 70 countries located in Africa, Southeast Asia, South Asia, and Latin America. These projects represent a remarkable 300 percent increase in hydropower development since 2008.[43] The reasons behind the recent outward orientation of China's dam construction industry are multifaceted. Support for overseas dam projects is as much a part of the Chinese government's current geopolitical vision—which is attentive to the international political economy of natural resources—as it was part of the United States' strategic planning during the Cold War. Hydroelectric schemes supported through Chinese funding mechanisms are frequently linked directly to energy-intensive extractive mining operations that would ultimately help meet China's surging demand for natural resources. For example, the China Machinery Engineering Corporation's support (later withdrawn) for the Belinga hydropower project in Gabon was an integral element of a US$3.5 billion iron ore mining operation that included roads, transmission lines, and a modern port facility.[44] Furthermore, government agreements to finance water infrastructure in African or Asian countries are often designed in such as way to ensure that Chinese firms are given increased access to strategic resources, including oil. This was the case with Chinese support for the Merowe Dam in Sudan, which came as part of a package of trade and aid relations between the two countries involving oil exports to China and a host of other development projects to be implemented within Sudanese territory. Building dams overseas is clearly profitable for Chinese state-owned enterprises, which have become more "corporatized" in recent years, and the outward orientation of the dam industry has been strongly encouraged by the Chinese government through foreign investment incentives and other measures.[45]

Making any definitive statements on the precise nature of the Chinese state's geopolitical aims in disseminating technical and financial assistance for large dams would be unwise. Still, a growing body of research suggests that China's foreign assistance program is directly linked to a complex blend of government rationales: to expand overseas opportunities for Chinese companies, particularly in the engineering sector; to secure access to crucial energy and mineral resources to meet growing domestic demand; to promote Chinese aid as a more transparent and less conditional alternative to conventional Western aid (e.g., aid from OECD countries); and, flowing from the previous goal, to expand China's influence in world politics by generating geopolitical alliances among the states of the tricontinental world.[46] These factors coalesce around a strategic geopolitical vision that sees China's dam-building activities in Africa, Southeast Asia, South Asia, and other regions—often in countries ignored by traditional donors because of their autocratic regimes and maltreatment of their citizens—as part of the government's efforts to curry international sympathy, in part to deflect critical examinations of China's own human rights and environmental records, but also to build its own "Third Worldist" partnership in contradistinction to the United States and its geopolitical confederates.

Like their American counterparts of an earlier era, Chinese dam-building agencies and companies have built up a prodigious expanse of technical expertise and experience based on domestic water resource development that has contributed to the initiation, design, and construction over half of the world's 50,000 large dams. Yet unlike the Bureau, the Chinese enterprises are typically engaged in all aspects of dam building, from the reconnaissance and design phases through the construction of all or part of a hydropower project, and oftentimes even its operation. The prodigious financial resources of the Chinese state—often filtered through bilateral aid agencies such as the Export-Import Bank—provide funding for many of the projects that have engaged Chinese expertise. Chinese dam-building firms thus wield an extraordinary degree of power over the governance process that extends far beyond their knowledge of dam design and construction. What is not at all clear is the degree to which water resource experts and foreign policy architects within the Chinese state communicate and negotiate the geopolitical dimensions of their activities, although it is likely that expertise and geopolitical goals are tightly aligned, given the significant overlap of officials engaged in both realms.[47] At a slightly different level, Chinese companies and the Chinese government have also been centrally involved in designing and funding a number of projects—prominently in Laos and Ethiopia—that have exacerbated tensions over the sharing of transboundary waters. For example, China's domestic expansion of hydropower development—particularly in the upper reaches of the Mekong River basin in China's Yunnan Province—has generated concerns over its potential effects on the downstream regions of transboundary basins.[48]

It is difficult to say precisely how China's hegemonic influence on the current era of large-dam proliferation will shape the future of water-society relations. This hegemony is packaged within a highly effective message to states of the tricontinental world: allow us to finance and build dams and grant us preferential access to vital resources within your sovereign territory, and the benefits of hydroelectric development will accrue to your society. Everyone "wins" under this optimistic scenario, except, of course, the socioecological actors displaced or disrupted by river alteration. There are some signs that the Chinese state is internalizing the critiques of its dam-building programs—both domestically and abroad—and instituting reforms that attempt to mitigate the more severe social and ecological effects of hydropower projects.[49] Conceptually, the emergence of Chinese enterprises as "dam builders to the world" may call for a reworking of our view of the networks assembled through large-dam construction. Geopolitical aspirations are still critically important components of the technopolitics of large dams, but are perhaps not as significant as during the Cold War. Dams and technical assistance are no longer conceived as a "weapon" directed against a competing geopolitical and ideological opponent. They have become, instead, more subtle tools of influence, ones that offer potentially vast economic and geopolitical benefits to the states and companies promoting their dissemination as well to the governments of the territories where they are sited. Yet as the political ecology of large dams makes clear, these tools and their wielders are being increasingly challenged by an array of dam-affected communities, the transnational advocacy networks that support them, and environmental scientists concerned with the short- and long-term impacts of damming.

A "New" Political Ecology of Damming Rivers

The current technopolitical networks driving the construction of large dams throughout Asia, Africa, and Latin America constitute a fast-moving analytical target. While the projected plans of dam-exporting governments, such as China and Iran, and the professed goals of international financial institutions such as the World Bank in expanding economic resources for water resource development are clear, the precise character and locations of the hundreds of proposed projects and their specific socioecological impacts are ambiguous. The ambitious plans of project promoters almost always outpace the actual implementation of dams. This section shifts the analysis in some ways toward a political ecology of large dams in the hopes of exposing the fault lines of political conflict over water infrastructure, which reside at manifold institutional and societal scales and levels. A political ecology perspective emphasizes the myriad anthropogenic origins of the transformation of rivers through damming while giving equivalent analytical weight to the materiality of the biophysical and social transformations that inevitably accompany and follow dam construction.[50] A political ecology of large dams and river alteration in the present era, as the preceding sections highlight, must contend with an array of novel geopolitical arrangements, assumptions, and knowledge regarding river systems and the political struggles arising at several spatial scales over the genesis and implementation of large-scale water development schemes.

There are three dimensions of the global debate over large dams that are different from, yet related to, the facets highlighted in the book's previous chapters: an effective and diverse global anti-dam social movement; debates over hydropower as a "clean" form of renewable energy vis-à-vis climate change; and an emerging scientific understanding of rivers, particularly large rivers, as complex systems. Taken together, these aspects of the political ecology of large dams represent, to a certain degree, novel networks of knowledge and social action. Seen in light of the histories presented here, they also contribute to an updated understanding of the networks that are assembled by large dams (see chap. 1). In some sense this assemblage of networks highlights the prominence of discourses of hydropower that bring together an impressive array of things and processes—technoscientific, socioecological, symbolic, financial, and so on—that are characteristic of the contemporary era of large-dam geopolitics.

Emerging from broader critiques of development initiatives arising from a circumstantial coalition of scientific and activist groups in the 1960s and 1970s, there now exists a mature global movement focused on problematizing the economic rationales and socioecological effects of large dams. While this is not the place for a comprehensive history of the self-described "anti-dam" movement, I will sketch the contours of its evolution over the past three to four decades and its main tenets. This movement connects temporally to the cogent appraisals of large dams in the United States in the late 1960s[51] as well as to some critical assessments of large dams as international development projects (such as Pa Mong, mentioned in chapter 5).

Despite the characterization of large dams as "careless technologies" that often produce more social and environmental harm than societal benefits, large dams continued to be linked to North-South programs of development assistance throughout the 1970s and 1980s. The Nordic countries (particularly Norway, Sweden, and Finland) were particularly aggressive in recommending hydropower development as a source of renewable, relatively inexpensive energy to governments of the Global South. Aid for hydropower development was linked directly to the Nordic states' successes in developing their own rivers into electricity-generating systems via large dams, and quite frequently such aid was accompanied by well-established firms from these nations undertaking actual construction of the projects, as, for example, in Southeast Asia.[52] Moreover, the Indian and Chinese governments, as well as several Southeast Asian states, embarked on remarkable dam-building programs throughout the period immediately following the cessation of the Bureau's overseas activities in the early 1970s. These programs resulted in a number of megaprojects, some of which (at least in India) prompted local and national resistance and achieved international notoriety, thus setting the stage for the global anti-dam movement and, ultimately, the convocation of the World Commission on Dams (WCD) in the late 1990s.

Two dam projects in particular, the Sardar Sarovar Dam on the Narmada River in India and the Pak Mun Dam on the Mun River (a tributary of the Mekong) in Northeast Thailand, seemed to catalyze the global anti-dam movement and allow for the rapid dissemination of experiences to dam-affected communities throughout South and Central America, Africa, South and Southeast Asia, and other regions. Both the Narmada and Pak Mun projects were first proposed by their governments many years before their scheduled implementation in the early 1990s, and both received support from the World Bank, although the bank withdrew its funding from the Sardar Sarovar project in 1994.[53] Although both schemes were eventually constructed, they gave rise to several NGOs that have since been very active in generating a global coalition of communities that have been displaced or otherwise disrupted by large-dam construction and other forms of river intervention. This coalition, which culminated with the First International Meeting of People Affected by Dams in Curitiba, Brazil, in March 1997, is notable for its inclusion of a wide range of spokespeople from communities actually experiencing dam-related disruptions, in addition to their affiliates in advocacy and educational groups. At the same time, a growing body of scientific knowledge on the cumulative biophysical and social impacts of large dams (and dams in general) was being absorbed, summarized, and circulated by international environmental organizations connected to more localized anti-dam campaigns in cases such as Narmada and Pak Mun.[54] The combined critiques of environmental scientists, global advocacy groups (e.g., the International Rivers Network, now simply International Rivers), and the coalition of dam-affected communities culminated in the creation of the World Commission on Dams (WCD) in the late 1990s and publication of its landmark report on global water governance in 2000.[55] As noted in chapter 1, the WCD's recommendations—revolving primarily around the need for more inclusivity and transparency in water development decision-making processes and options for delivering the benefits provided by large dams through less deleterious means—were greeted with a mixture of contempt and disbelief by the global dam industry and prominent dam-building governments such as China and India. At this point, the influence (if any) of the WCD and its report is largely undecided,[56] although the report does provide a powerful set of talking points for the global anti-dam coalition. If the state-initiated dam-building plans delineated earlier in this chapter are any indication, the WCD guidelines are unlikely to take precedence over the more overt political actions that have been one of the hallmarks of communities around the world protesting large dams. Ultimately, the significance of the global anti-dam movement resides in its continuous questioning of the deleterious impacts of large dams from a position that links ecological change to questions of livelihoods and social justice. The thousands of communities across the planet that have been and are being affected (and threatened) by large dams have become firmly enrolled within the assemblage of networks drawn together by large dams.

A second critical dimension of the contemporary political ecology of large dams concerns recent scientific debates about the contributions of dams to carbon emissions and hence to global warming. In essence, the debate—mostly focused on tropical regions, but relevant to temperate zones as well—boils down to whether the reservoirs of large dams constitute a significant source of emissions of methane and carbon.[57] One of the earliest studies (reservoir emissions were not researched before the early 1990s) found that reservoirs are potentially significant contributors to planetary greenhouse gas fluxes because they eliminate carbon sinks (i.e., the inundated plant material, especially in forested areas) and release greenhouse gases to the atmosphere as the carbon in soils and sub-merged organic material is broken down and converted to carbon dioxide and methane.[58] The rapidly evolving science of reservoir emissions encompasses hitherto unexamined aspects of the biogeochemistry of reservoirs and hydroelectric production.[59] However, one of the central premises of the most recent push toward hydropower development—as evidenced by the ambitious dam-building programs of a variety of states and the rhetorical commitment of international funding institutions such as the World Bank to magnify support for dam construction—is the assurance that hydroelectric dams are a clean source of energy vastly superior to fossil fuel combustion. In this instance, scientific knowledge about the greenhouse gas emissions of reservoirs is being incorporated into the orbit of the large-dam debate, but in ways—like the now decades-old understanding of the array of biophysical disruptions accruing to dammed rivers—that are accorded secondary status. Perhaps most importantly in the context of this book, the nexus of hydropower and climate change is now part of the discursive landscape of the geopolitics of large dams, something that would have been unimaginable in the time of the Bureau's work. This, again, accentuates the resilience of the technopolitical networks constructed around large dams and basin development of a certain kind. The resilience and adaptability of these networks, originally catalyzed by the brute economic and developmental arguments of dams of the past, is exemplified by their capacity to incorporate an environmentalist discourse on renewable energy that undergirds and provides additional rationales for the large dams of the present and future.

The political-ecological ramifications of large dams are also being shaped by contemporary understandings of large river systems as complex entities defined by a host of anthropogenic, geological, hydrological, and ecological processes, as well as by how we think about the materiality of nature in relation to social processes. On this latter point, the diversity of entities too often generalized as "resources," and the quite divergent characteristics of those "resources" (e.g., fish versus forests versus minerals versus genes), generate quite different relations with human social processes and meanings.[60] Thinking about dams in this light complicates our understanding of their functions and meanings. For instance, it is impossible to analyze dams—their conception, design, implementation, and consequences—without simultaneously analyzing and accounting for the unique characteristics of water and, moreover, the dynamics of water within a river system. As we have seen in many instances throughout the previous chapters, the river *basin* was perceived as the ideal vehicle for understanding a river system and its land-water interactions. However, the basin ideal was always predicated on developing and otherwise manipulating a river's flows, regardless of how such interventions might alter a river system's coupled hydrological and ecological dynamics. This perception has certainly changed in the last several decades as approaches to river governance grounded in integrated water resource management (IWRM) and adaptive management have stressed greater attention to the many ecosystem services provided by unaltered river flows. Although criticized as yet another politically naïve "Nirvana concept" within the field of water resources management, IWRM emerged in the 1990s as an alternative way for water managers and water policy decision makers to think about how to balance the multiple and often competing uses of freshwater resources among diverse stakeholders.[61] It is also clearly linked to a basin-oriented approach to water use and management. However, in contrast to the emphasis on the basin as an economic unit in previous eras, IWRM endeavors to include the hydrological and ecological dimensions of flowing rivers—and their benefits to human societies—in the calculus of water governance. Yet approaches under the IWRM rubric also allow for the integration of large dams and their multitude of effects within its management purview.

In a different scientific vein, recent biophysical research on the combined spatial and temporal dynamics of river systems—particularly large river systems—provides a portrait of the aspect of river basins that has ironically been the least understood: the biophysical processes (e.g., geological, hydrological, ecological) that define basins as material entities aside from human interventions. In short, a large river system exhibits an exceptionally complex and variable array of hydrological and ecological relationships across space and time that make it possible to describe it as an integrated system, but also make it especially difficult to predict the consequences of human interventions in this system.[62] To give slightly more detail, river systems are characterized by a high degree of spatial variation (e.g., hydrogeomorphic "patches" of different sizes) that is directly related to the temporal variation of system processes (e.g., changes in annual flow rates, varying flood pulses). While the precise implications of this novel understanding of river systems are not yet well understood, it is clear that large dams "impose an environmental homogeneity across broad geographic scales" that will undoubtedly lead to a global reduction in the biological diversity of riverine systems as more projects are built.[63] While there are myriad ideas and management approaches for mitigating the homogenizing effects of dams,[64] there is little evidence that the scientific knowledge base coalescing around large river systems—a knowledge base grounded in fluvial geomorphology, hydrology, and aquatic ecology—is being incorporated in any meaningful way into the decision-making apparatuses of the national and global actors promoting enhanced hydropower development. The current assemblage of networks that circulates around large dams—which both shapes and is shaped by an array of economically and politically powerful actors—is highly selective in the technical and scientific expertise that is mobilized to foster hydropower development. Climate change science and its (contested) prescriptions for generating "clean" energy mesh well with a discourse of accelerated dam construction, while the equally compelling science of large rivers is only partially enrolled within this particular technopolitical network. Determining which science counts and receives technopolitical support is a highly selective process.

Conclusion

My brief foray into the political ecology of large dams and river basin transformation serves as a fitting conclusion to this chapter by bringing a sharper focus to the numerous forces shaping the geopolitics of large dams in the present era. In comparison with the connections between massive hydropower projects and Cold War geopolitics, the technopolitical networks that define and drive (and are constituted by) large dams at present are more diverse and more complicated, yet are also remarkably familiar. Current water development planning in Lebanon, Ethiopia, and mainland Southeast Asia retains tangible technical and ideological connections to the Bureau's engagements in those places throughout the 1950s and 1960s (explored in earlier chapters). The Bureau's near-monopoly on the technical expertise required to undertake large-dam construction and river basin development has ended, and a powerful collection of Chinese companies and agencies—with strong links to the financial and political resources of the Chinese state—have emerged as central conduits for the dissemination of dam-related knowledge and dam construction.

What is also clear from this political-ecological approach to large dams is that deployments of scientific knowledge are now more diverse, but remain contentious. The debate is, however, far more visible to the public eye today, in sharp contrast to the Cold War era, when debate over large dams was virtually nonexistent and their geopolitical roots were largely hidden. The geopolitical motivations and rationales that drove the "concrete revolution" during the Cold War era must now contend with coalitions of dam-affected communities, advocacy organizations operating at national and global levels, and a range of scientific actors that question the continuing efficacy of large dams in light of their often severe socioecological consequences. As the concluding chapter explores, this shift creates institutional and political spaces for rethinking the geopolitics of large dams and generating technopolitical networks that are not solely at their service.

SEVEN

Conclusion: Large Dams and Other Things

Emphasis on and progress in technology generally has outstripped progress in sociology. We have learned how to do marvelous things with materials but we have not learned enough about fundamental human desires, values, and objectives nor how to attain them. In other words, we haven't learned how to apply our vast technical advances to meet the basic values and desires of people.[1]

—Gilbert Stamm, Assistant Commissioner, Bureau of Reclamation, 1969

Perhaps no other technological object has the ability to capture and enroll within its orbit as many biophysical, technological, political, economic, and ideological processes and things as large dams do. Yet this capacity to assemble often carries with it significant social and ecological ills. Gilbert Stamm seemed to sense this dichotomy. Stamm's statement above is a remarkable admission for the leader of an organization whose stated mission was (and remains) to improve human well-being through the judicious application of water resource technologies. The time of his confession of disquiet about technology, the late 1960s, was a transitional period within the historical arc of large dams and river basin development. As noted previously, the Bureau's domestic program was increasingly coming under assault from environmentalist critics (Stamm calls out the "sociologist, anthropologist, archeologist, geographer, and historian who stand on the sidelines offering negative criticism of plans for wealth-generating . . . development programs") who decried the inundation of lands of tremendous ecological value, especially in the American West. Globally, large dams were facing increasing scrutiny not only because of the undesirable ecological transformations they brought about, but also because of the myriad social disruptions—ranging from outright displacement to cultural dislocation and increased health risks—fomented by their implementation. As the previous chapter shows, these criticisms have only become more amplified and prevalent in recent years. Yet rivers continue to be dammed.

One of this book's central arguments is that the history of large dams, and of river basin development more generally, is simultaneously environmental, social, technical, and geopolitical.[2] My intention throughout has been to highlight the importance of the geopolitical dynamics of the Cold War in facilitating an alignment of economic and technical networks of development highly favorable to the dissemination of knowledge and ideological rationales surrounding large dams. This was a critical moment in the genealogy of altered rivers. This configuration of geopolitical and technological networks, while not the sole factor in explaining the twentieth century's "concrete revolution," provided crucial impetus for damming rivers on a planetary scale and found a willing and highly capable vehicle in the Bureau of Reclamation, a dam-building agency situated within the state apparatus of a global hegemon. Once the Bureau was "unleashed" from its domestic operations by the changing landscape of foreign assistance and development aid in the early 1950s, its technical experts and bureaucratic experience found willing hosts in the multitude of newly independent, developmental states throughout Asia, Africa, the Middle East, and Latin America. I have stressed Bureau engagements in China, Lebanon, Ethiopia, and the Mekong region, but given the remarkable scope of the Bureau's international activities, these examples could have readily been replaced with others from Turkey, Liberia, Iran, and the Philippines (see the appendix). The cases featured in this book, however, all demonstrate that the seemingly straightforward aim of the Bureau's overseas technical assistance program—laying the groundwork for water resource development and the modernization of river basins—was always complicated by institutional conflicts and the materialities of specific places. Ranging from tensions over technical assistance versus geopolitical strategies within the American state to the numerous administrative, political, and environmental problems on site, the technopolitical networks that congealed around dams and river basin development were always tangled. Hopefully, disclosure of these snarled political, technological, and ultimately environmental histories involving the Bureau, the American state's foreign policy apparatus, an array of developmental states, and the water projects they yearned to build and the rivers they expected to harness, reveal segments of geopolitical history that promote a reconsideration of the ultimate origins of large dams and river basin development. The experiences of the Bureau of Reclamation and its role in the geopolitics of Cold War–era development may also prompt a rethinking of technopolitics and the efficacy of large dams more generally.

This book will accomplish an additional goal if it provides fodder for reinterpreting the evolution and meaning of large dams in the context of both scholarly and public debate. It is comforting to the public and, indeed, to many policy makers to think of all aspects of large dams—their location, their design and construction, their justifications—as purely technical decisions. Surely things as large and potentially catastrophic (if they fail) must be beyond any political-economic calculations. The evidence presented in this book refutes this claim. This is not to say that large dams are not based on technical decisions, or that dams as a technology have some particular claim to being exclusively entangled with politics. Rather, my simple argument is that all dams, as is the case with any technological extension of society, are the result of technopolitical decisions and conditions. Thus I am hopeful that the histories and analyses presented here contribute to scholarly endeavors. For example, unearthing the manner in which geopolitical dynamics intersect with, transform, and on occasion are transformed by, technical expertise remains a crucial task for scholars of science and technology studies, critical geopolitics, environmental history, political ecology, and other fields interested in the co-production of water, politics, and technical knowledge. More pragmatically and politically, efforts to reformulate the forces that constitute, for example, the Mekong basin as a dammable river system—whether grounded in calls for ecological and livelihood sustainability or justice for the human and nonhuman entities facing violent disruption under currently proposed development programs—would, I hope, benefit from historical thinking about the specific technical expertise and geopolitical practices directing water governance. I examine this theme more fully below, but turn now to a more detailed consideration of the conceptual dimensions of the geopolitics of large dams featured in this book.

Dams and Technopolitics

Gilbert Stamm's juxtaposition of "marvelous things" with attainment of "human desires" represents a pointed crystallization of the curious mixture of development aspirations and political clear-headedness that undergirded the Bureau's overseas operations during the Cold War. At a conceptual level (and as insinuated above), all technical assistance, and hence the knowledge and people that prop it up and make it feasible, is inherently political. Yet this rather banal statement is something of an epistemological cop-out. It calls for further refinement. The technical knowledge wielded by Bureau engineers and other experts co-produced a social order deeply imbued with geopolitics.[3] This co-production was incubated within the dynamics of what

I have referred to throughout this book as technopolitical networks, and it received contributions every time a Bureau staff member recorded a flow rate, analyzed a soil sample, projected energy demand, or designed a dam, whether such an act was carried out in Chinese, Lebanese, Ethiopian, or Thai territory. It also occurred every time an official within the US State Department dictated a memorandum, offered geopolitical judgment, or made a decision about how to distribute American development assistance. The outcome of these thousands of contributions was a technopolitical order, one that was simultaneously specific to each of the sites of Bureau intervention and generative of the broader geopolitical goals of American hegemony. It is these details of the construction of technopolitical networks that I have brought to light, and if this book accomplishes nothing else it will at least, I contend, add to our collective knowledge of a significant yet largely concealed element of the Cold War's ideological and environmental legacies.

Additionally, the technologies disseminated throughout the tricontinental world during the Cold War under the rubric of foreign and technical assistance manifest a special kind of technopolitics, one that was patently an outcome of geopolitical calculations. Previous studies examining the intersection of technology and politics in the context of development programs demonstrate brilliantly the complex interplay of state ideologies and specific historical-geographical contingencies that have propelled forward profound societal transformations under the auspices of "modernization."[4] Yet such studies have left unexamined, or at least relegated to a secondary role, the broader geopolitical forces that drove so much of the international development agenda during the twentieth century. To be fair, I have certainly been derelict in not digging deeper into the local and national political ecologies and political economies of the places that experienced Cold War-inflected technical assistance. My goal has been to shift the analytical lens toward the processes and actors involved in constructing the technopolitical networks of water resource development. Disclosure of the geopolitics of technical assistance, as filtered through the myriad activities of Bureau water experts, their engagements with material, institutional, and political landscapes in various parts of the world, and their negotiations with foreign policy apparatuses also demystifies and leaves less abstract the conduct of geopolitics. A focus on the Bureau reveals more clearly a critical agent of the geopolitics of development—the "technical expert"—and an associated bureaucratic environment, albeit an agent that enacted geopolitical objectives in often unexpected ways.

One of the salient epistemological challenges of critical geography is paying attention to the ways in which various spatial scales are constructed through discourse and practice. Scalar thinking—always filtered through powerful imagined geographies—can contribute to an understanding of technopolitics in several ways.[5] At one level, the technopolitical networks that circulate around large dams—certainly around the ones described here and almost surely around all large-scale technological projects—are given motive force by imagined geographies pegged to certain scalar configurations, or scale-making projects. The architects of American foreign policy, as evidenced throughout this book, perceived a "third" world—as yet outside the realm of Soviet influence—of economic immaturity, untapped resources, and susceptibility to communist political ideologies. And they placed this world on a map and labeled it "underdeveloped." The Bureau's technical expertise related to dams and water resource development was mobilized in order to bring this unruly global space within the orbit of American influence.[6] This space was and is occupied by the category of the nation, a scalar categorization that constitutes the central territorial entity of world politics. In terms of Cold War technopolitics and large dams, the nation—a constructed space—served as the fulcrum of developmentalism: despite their vast differences in culture, political institutions, and geographies, American technical expertise and its geopolitical accouterments tended to perceive the China of the 1940s and the Lebanon, Ethiopia, and the Lower Mekong countries of the 1950s and 1960s as commensurate in their desire for modernization and their susceptibility to communist influence. This perception, of course, reinforced the ways in which the world's political geographies were imagined, classified, and treated by the Cold War architects of foreign policy and their technical assistance proxies. In a similar way, the scale of the river basin served as an organizing spatial unit for nearly all of the Bureau's water resource activities, and as previous chapters point out, it also served a homogenizing function that resulted in remarkably similar technical approaches to water resource development worldwide. And, finally, the scale of the dam site itself—the most direct focal point of the Bureau's Cold War intercessions—constitutes a crucially important scalar configuration by bringing development expertise and ideologies into contact with the messy contingencies of *place*. A comprehensive accounting of technopolitics must follow the agents of development and geopolitics as they traverse, produce, and give institutional salience to these multiple scalar categories. Moreover, while virtually all development technologies can be characterized as multiscalar, certain types of technologies—large dams, for example—are distinct in their capacity to draw together and arrange technopolitical networks across significant distances and durations. One of the benefits of scaling technopolitics in the way that is sketched here is that it allows for an analysis of power relations—whether these be disposed through economic, political, social, environmental, or hybrid constellations of networks—as they cut across times and spaces.

As the foregoing discussion demonstrates, a conceptual framework emphasizing how specific technologies were historically embraced and transformed in unexpected ways by geopolitical processes is equally relevant to the analysis of present-day development interventions. An understanding of technopolitics is particularly important when so many of these interventions, now as in the past, are presented by backers in governments and aid agencies as simple exercises in rational decision making and scientific management. The genealogy of large dams as geopolitical objects—constituted through and constituting multiple temporalities (see chap. 6)—helps make visible the ideological roots of large dams (and indeed, any seemingly neutral technology) and brings to light the now hegemonic idea that large dams and river basin development are the preferred means of organizing water-society relations.

A corollary goal of this book has been to trace the powerful configurations, or networks, that explain the potent capacity of dams to transform landscapes, peoples, and national economies, both materially and in terms of how we perceive such altered entities. My focus on a specific agency, the United States Bureau of Reclamation, undoubtedly ignores important dimensions of the genealogy of the proliferation of large dams and the river basin ideal. However, a careful accounting of the Bureau's activities as a crucial agent of dam building and river basin development—based on the largely opaque records documenting communications concerning the strategic importance of water development as it occurred within the foreign policy apparatus of arguably the world's most powerful state in the twentieth century—exposes something important in the realms of environmental history and the geopolitics of development. Within a broad range of social and human sciences, there is now a growing awareness that in order to understand environmental changes and their effects on social relations, one has to simultaneously explain the manner in which environmental transformations are both socially mediated and ecologically constructed.[7] An important part of these constructions is the influence of knowledge production, particularly scientific and technological knowledge production.[8] Moreover, the twentieth-century diffusion of large dams—a concrete revolution—clearly demonstrates the ways in which particular technologies and forms of knowledge are both integrated within networks of political calculation and generate broader networks of expertise, ecological relations, and in this case, geopolitics. At a basic level, my goal has been to shed light on the "extra-scientific origins" of so many of the technological interventions carried out in the name of economic development over the second half of the twentieth century and on how these origins reverberate in the present day.[9]

The Mekong River basin project in particular shows why explication of these themes—the globalization of large dams and river basin planning, and their connections to geopolitical and technological thinking—is critical for understanding present-day concerns over river regulation. As related in chapter 5, the early years of the Mekong project (the late 1950s through the mid-1960s) were dominated by hydrological investigations and reconnaissance studies to identify likely tributary and mainstream dam projects. Throughout this period the cooperative development of the basin was invoked as a means of bringing peaceful relations to the region. The Bureau, through its work on the Pa Mong Dam, became a critical element of the technopolitical network that was coalescing around Mekong development. But—and this brings us to another important notion—the network of expertise and geopolitical relations outlasted the technology (i.e., the Pa Mong project itself) that, at least in part, brought it into being. Seen in this light, the project surely never accomplished any of its intended outcomes: damming of the river for irrigation development, the production of hydroelectricity, and hence industrialization; serving as the linchpin that would lead directly to exploitation of the entire basin; or ushering in an era of peaceful relations based on cooperative basin development among the nation-states of a "shatterbelt" region being ripped apart in the name of more powerful Cold War governments.

Indeed, if we add to Pa Mong the other cases presented in these pages, the promotion of large dams and river basin planning throughout the tricontinental world as a geopolitical tool for containing the spread of communist ideology emerges as slightly more than a symbolic effort with little to show in terms of strategic advantages, yet certainly a great deal less than an entirely successful initiative that achieved its geopolitical goals. But this is not the point: the lasting legacy of these projects is not their collective geopolitical effects, but rather their far more substantial impacts on livelihoods and landscapes. Analysis of the Bureau's "foreign activities" thus confirms, albeit in a narrower and more detailed way, James Ferguson's argument that we must interpret the "success" or "failure" of development projects in ways that de-emphasize the original intentions of the interventions and instead connect their outcomes to a "kind of political intelligibility" that at first may not be apparent.[10] In this instance, the broader constellation of power and influence that the development of the Yangtze, the Mekong, the Litani, and the Blue Nile corresponded to—its "political intelligibility"—was the global circulation of the ideas and practices that perceived large dams and river basin development as a universal "fix" for water resource development. So while nearly all the projects investigated here failed in geopolitical and developmental terms, they all became part of the expanding work of the Bureau of Reclamation and the particular set of technologies and ideologies (the technopolitics) built around exploiting water in certain ways. And the Bureau was an almost ideal American vehicle for abetting this proliferation. Moreover, Cold War geopolitical conditions virtually assured that this transfer of technology and accompanying ideology would be, in theory, relatively smooth. The involvement of the Bureau created depoliticized spaces from which to initiate and pursue development projects (e.g., dams) and approaches (e.g., basin-oriented development) that Bureau staff were ideally trained to disseminate. Perhaps most importantly, the creation of these spaces allowed discourses of river basin development and hydropower production to more readily transcend political boundaries and become globally recognized and—at least in the view of states and their confederates in financial institutions and the private sector—desirable phenomena. Yet, to borrow Anna Tsing's evocative term, there was tremendous "friction" where this universalized and globalized notion encountered entrenched political-economic interests and nature's materialities in distant lands, and that friction nearly always confounded these projects' original intentions.[11]

Water for Peace?

In May 1967 a significant yet largely forgotten convention took place in Washington, DC. The International Conference on Water for Peace brought together about 1,200 official delegates and 2,800 observers (encompassing government officials, water engineers, development planners, journalists, and others) from over 100 nations for a series of speeches, papers, and plenary sessions. The convention was a direct outgrowth of Lyndon Johnson's initiatives in the Mekong basin (see chap. 5) as well as a growing sentiment within US foreign policy circles that water, particularly the water flowing through river basins shared by more than one country, could be a powerful symbol of international cooperation and economic development. At the closing ceremony, Secretary of State Dean Rusk spoke eloquently of the "unifying" power of water. After detailing the challenges associated with water resource management in terms of the "most sophisticated techniques of scientist, engineer, administrator and educator," Rusk confessed he was "better equipped" to address water's "influence in foreign affairs."[12] "Water," argued Rusk, "tends to unify, and not to divide" and "normally facilitates the arts of peace. It causes agriculture to flourish, and turns the wheels of industry and commerce. The availability of plentiful supplies of water is not likely to direct a nation's thoughts to aggression—rather the reverse." He proceeded to make the case that water's capacity to foment peaceful societal relations is universal, noting that

> we see the same unifying principle operating upon practical measure. . . . Slaking the thirst of arid regions by converting brine to fresh water involves the same industrial processes, whether that region is in the Near East or the American Southwest. . . . Erecting dams, controlling floods, sinking wells, collecting hydrological data: all these activities are carried out in much the same way, regardless of location. The techniques of water resource management are to a considerable degree applicable everywhere, therefore are transferable.[13]

There is much to take umbrage with in Rusk's rhetoric. But for the moment, I want to set aside justified critiques of the technical interventions he mentions that have proved ineffective and even harmful, and of his advocacy for a single management approach to what are vastly different waterscapes and rivers residing in different historical-geographical contexts. Instead, I want to argue that there are kernels of a radical project in Rusk's linkage of water and peace, albeit one far removed from the diplomatic and political focal points that provoked his remarks. Somewhere buried within the institutional structure of the Water for Peace Conference and the subsequent creation of the short-lived Office of Water for Peace within the US State Department is an alternative ideal, one that is potentially counterhegemonic in terms of confronting the renewed calls for large dams and hydropower development featured in chapter 6: What if we were to take the potential relations—whether political, technical, cultural, or all of the above—between water and peace seriously?

These relations underscore the hitherto somewhat obscure normative aims of this book. My argument in this regard is simple: large dams and river basin development—and technologies in general—are not "the problem" if one is concerned about the socioecological disruptions perpetrated in the name of capitalism, totalitarianism, "development," "progress," or any of the dozen or so other signifiers of material and social relations that have contributed immensely to adverse conditions for the earth's human and nonhuman inhabitants. The problem lies instead within the technological and geopolitical (and political-economic) networks that sustain the idea that large dams and river basin development are the most effective ways to structure water-society relations for the betterment of humanity. In essence, I am asking whether the assemblage of networks that produce and maintain large dams—materially and ideologically—can be reconfigured to produce and maintain something else. This "something else" might still include large-scale water infrastructure, subject to discussion of alternative methods of providing the benefits offered by dams. Surely the extreme social disruptions visited upon dam-affected communities over the course of the "concrete revolution" should be condemned and can be avoided, but as the experience of the WCD makes clear, hydropower proponents in governments, financial institutions, and the water development industry perceive such disruptions as an unfortunate trade-off against the "greater good" of economic development. The networks that large dams assemble do not allow for an alternative way of thinking.

How, then, would a technopolitics of "water for peace" take shape? Perhaps it would not be a network composed of technical expertise and geopolitical ambitions of the sort described in these pages. Instead, it might embrace networks of a fundamentally different type with different kinds of attributes. More than anything, a connection between peace and water implies more inclusive and deliberative processes of water governance. Currently these processes are variously supported, hampered, undervalued, and co-opted by the variety of political institutions charged with making decisions about how societies relate to water. Furthermore, the assemblages of networks that produce and maintain large dams are remarkably undemocratic. Representatives of the humans and nonhumans that are deeply affected, often in negative ways, by water resource development are largely ignored in the decision-making processes that result in the construction of a large dam. Such decisions are instead an outcome of technopolitical networks forged in large part by states and their technical experts and the enormous financial resources and political influence required in order to even contemplate the construction of large-scale projects. But these are disempowering observations, tending to lead to the conclusion that the networks assembled by large dams are virtually impregnable. I thus want to conclude with a story demonstrating how different networks might be assembled (or reassembled) in ways that connect the idea of water for peace with large dams and their technopolitical networks.

In February 2010 an array of social actors that had been engaged in a long conflict over water rights in a highly altered river basin signed the Klamath Hydroelectric Settlement Agreement, a component of the broader Klamath River Basin Agreement, a covenant that called for the removal of four major

multipurpose dams distributed along the Klamath River in northern California and southern Oregon to achieve an assortment of river restoration
goals. None of the dams have been removed to date, and their decommis-
sioning faces significant hurdles,[14] but the likelihood of what will eventually
be the "largest dam removal project in history" is noteworthy at many levels.
Most noteworthy is that the removals were negotiated and agreed to by a
prominent utility company, an assortment of American Indian tribes, water
user associations representing landowners benefiting from the dams, environmental organizations, and a variety of other actors, a coalition whose
"stakeholder interactions" had historically been characterized by "animosity
and distrust" for decades. A prominent part of the vision of the agreement,
inscribed in its text, was the aim "to achieve peace on the river and end
conflict."[15] This is a far cry from the type of water-related peace imagined by
Dean Rusk, and it presents an intriguing, alternative way of thinking about
water-society relations and the technopolitical networks that sustain those
relations toward certain ends.

Indeed, the contrast is stark. Rusk was giving voice to a geopolitical
imagination that perceived the damming of rivers as a path toward cooperation among nation-states, which would presumably reap the national
economic benefits of developing their water resources, particularly in cases
in which those waters were shared across national boundaries. Peace within
this configuration of geopolitics and technology was almost purely at the
level of state-state interactions; the violence and conflict surrounding the
implementation of so many of the world's large dams—centered on displaced communities, lost livelihoods, and degraded riverscapes—was an
afterthought. In contrast, the political, cultural, biophysical, economic, and
technical processes and relations coalescing around dam removals on the
Klamath and elsewhere in the world present peace and water in a vastly
different light. Peace here is concentrated directly on the river and its basin
and on the array of human and nonhuman agents that see improved water-
society relations flowing from an undammed river.[16] In effect, a collection of
networks is being drawn together, or perhaps reassembled, around a hybrid
of river and human society that does not have a technological object as its
focal point. This would be a very different kind of revolution. We should
not be overly sanguine about the prospects of dam removal becoming so
commonplace as to counter the potent technopolitics driving the current
era of dam construction. But dam removal does present an alternative path.
I began this book with the contention that dams are lively entities, capable
of acting in unexpected ways. Rivers and water are lively as well and equally
capable of shaping the relations between humans and nonhumans, between

technologies and societies, and between technologies and natures in surprising ways. Redefining, or in some cases recovering, the assemblage of networks that characterize water-society relations and nudging them toward a condition of peace is a central challenge of the current century, and one well worth taking on.

Geographical Scope of Bureau of Reclamation Activities, 1933–1975

This appendix illustrates the broad range of the Bureau's overseas activities during the period that is the focus of this book. The cases herein are all examples of the Bureau's international engagements that go beyond the scope of this book, but are certainly deserving of expanded investigation and analysis. The examples I have selected do not include the Bureau's many interactions with water resource development agencies and experts from the states of the so-called developed world (e.g., Australia, Japan) because, as explained in the previous pages, a crucial component of the Bureau's geopolitically motivated technical assistance was to focus on "underdeveloped" regions.[1] My hope is that others will take up the technological and geopolitical threads presented here. In each entry, I offer a brief description of the Bureau's activities surrounding a particular dam project or river basin development scheme in which it had some level of involvement. In addition to summarizing the contributions of either the Bureau or other technopolitical agents in particular locales, I feature the (geo)political arguments constructed around each intervention in water resource development. Moreover, I have tried to be specific where specificity demonstrates the technopolitical character of Bureau and State Department work. These appendix entries should not be considered comprehensive, but rather as a complement to the broader themes of the book.

There is no obvious or "correct" way to organize the Bureau's engagements in non-US countries throughout the course of the twentieth century. I have consolidated these activities geographically, by world region and then by nation-state, although astute readers will recognize these entities as politically constructed.[2] Still, they do reflect how the US State Department, and hence the Bureau of Reclamation, organized the globe according to preconceived notions of geography, ethnicity, political organization, and so on. I have also included countries that, while technically not host to a Bureau team for even a short period, were deemed to be strategically important by State Department officials. Typically, this resulted in US embassy officials within the host country tracking and documenting its water resource and energy development—particularly hydroelectric development. This attention was partly motivated by the possibility of American business interests being able to capitalize on information and political conditions conducive to obtaining design and construction contracts. I have therefore included places and projects in which the US government was directly or indirectly engaged in the planning, financing, or construction of large dams without any significant Bureau involvement, and which were deemed strategically important (rightly or wrongly) by officials connected to the foreign policy apparatus of the American state.

Africa

Democratic Republic of the Congo/Zaire

The Bureau never played an active role in development of the Congo River system, but the United States—via both government and private channels—has had a sustained interest in the hydroelectric potential of Africa's largest river, especially within the framework of the Inga hydroelectric project. Throughout the 1950s and 1960s, Americans affiliated with a range of prominent water resource development agencies and firms—including the TVA, Morrison Knudsen, and Harza Engineering—expressed interest in the design, financing, and construction of a projected four large dams (two are finished) on the Congo River and associated transmission infrastructure, including the Inga-Shaba power line.[3] In the late 1960s the State Department monitored the discussions over financing of the proposed Grand Inga scheme, since both USAID and the World Bank were considered potential funders.[4] Today, development attention in the Congo River basin is focused on the astonishing Grand Inga Dam, which, if completed, would produce an estimated 39,000 megawatts of electricity and provide the hub of a vast energy network extending throughout the African continent and even to Spain and the Middle East.[5]

Ghana

Aside from the Aswan High Dam, the Volta River project is perhaps the best-known river development project in Africa. As recounted in several excellent studies, the United States was heavily involved in promoting the construction of the Akosombo Dam because it reflected the commercial interests of Kaiser

Aluminum and was a means of fomenting political linkages with the newly independent regime of Kwame Nkrumah in the late 1950s and early 1960s.[6] Somewhat unusual in the Bureau's history of foreign engagements was a study it undertook in 1963 on the resettlement process after construction of the dam.[7] The resettlement of approximately 72,000 people was funded in large part by a range of Ghanaian government agencies, and the government perceived that technical assistance and mass education delivered by these external agencies would be crucial to the success of the resettlement. Although the details of Ghana's river basin development program have been widely studied, the Bureau's report is notable for its recognition of the social disruptions caused by large-dam construction (at a time well before such issues were typically considered by water resource experts) and for its firsthand accounts of Ghanaian peasants forced to move. For example, the study highlighted the ways in which traditional land tenure and authority structures could complicate resettlement plans, as well as how government restrictions and actions could inhibit natural resource use and undermine livelihoods. Villagers responded with confusion to some basic questions such as "Why is it necessary that we should move?" The effects, if any, of the Bureau study on how the resettlement program progressed are not clear.

Ivory Coast

The United States did not contribute direct technical assistance to dam building in the Ivory Coast, but the American state employed funding for proposed water development projects as a geopolitical instrument. In the early 1960s the Kossou Dam, slated to be built on the Bandama River, was touted by Ivory Coast president Félix Houphouet-Boigny as a vital step toward boosting the economy of the newly independent country. It would ostensibly provide irrigation to help modernize farming methods, serve as a cheap power source for industry, and spur rural development.[8] In search of financial support, President Houphouet-Boigny requested financial assistance from the United States in the form of a US$40–50 million loan from the US Export-Import Bank, and in December 1965 President Johnson penned an encouraging letter to Houphouet-Boigny, assuring him that the United States would seriously consider providing support.[9] Recognizing the Ivorian president as "our most consistent and effective supporter in Tropical Africa," US officials maintained that a "failure to be forthcoming will be interpreted as a breach of faith by the Government of the Ivory Coast."[10] In a 1967 telegram, a US embassy official, while not wanting to be seen as imparting "Cassandra-like forebodings" of a Soviet takeover of the project, reminded State Department officials of the Aswan Dam debacle.[11] Running counter to the US government's political motivation to support the dam was the US Export-Import Bank's increasing concern over the Ivory Coast's ability to pay back the loan. The bank contended that the dam was likely to exceed original cost estimates, take over a decade to start earning money, and prove a hindrance to other economic development needs of the Ivory Coast.[12] Ultimately, despite the bank's misgivings, the American and Ivory Coast governments signed an agreement committing the bank to grant a US$36.5 million loan to the Ivory Coast for the financing of the dam.[13] The dam, which displaced 85,000 people, was completed in 1973.[14]

Kenya

In April and May 1967 the Bureau carried out a relatively swift reconnaissance evaluation of the Kano Plain project, an irrigation development program that would cover an estimated 30,000 acres in western Kenya. The three-person Bureau team concluded that the project was potentially viable, but that additional studies, expanded to include a more comprehensive survey of Kenya's portion of the Nile basin, should be carried out.[15] It was not immediately obvious why the Bureau needed to conduct a review of the Kano Plains project, since a British engineering firm had completed a feasibility study in the late 1950s, which showed the project to be sound on technical and economic grounds. However, USAID officials had received information early in 1966—from the British embassy in Washington—that the Kenyan government was anxious to begin the project because the "inhabitants of the area were expecting the scheme" and the "area was politically sensitive." Funding for the project was originally linked to an agreement with the Soviet Union signed in 1964, which required the Kenyan government to import Soviet consumer goods to cover the project's local costs in return for Soviet aid. When the Kenyan state informed the Soviets that this requirement was unacceptable, the Soviet aid apparatus offered a loan instead, which the Kenyans deemed overly exploitative. Kenyan officials were thus seeking an alternative funding source, apparently the United States via USAID, for the irrigation scheme.[16] This history places the Bureau's study clearly within a broader set of geopolitical motivations.

Liberia

The Bureau's undertakings in Liberia, which represent one of its earliest forays abroad, revolved around its Point Four–sponsored investigations in the early 1950s of the development potential of a moderately sized hydroelectric project, which eventually became the Mount Coffee Dam on the St. Paul River. While certainly not the most intensive of the Bureau's technical assistance efforts, the Liberia study nevertheless demonstrates the often tortuous route between project conception and completion and the unanticipated hindrances to seemingly straightforward technical assistance programs. In addition to a paucity of critical hydrological data, Robert Williams, head of the Bureau team, discovered serious misconceptions about his and the Bureau's roles upon entering the country. At one point he stressed to US embassy officials in Monrovia that he "was not to be considered as being out here to set up and head a Division of Water Control for the Liberian Government," which was apparently assumed to be the case by other American personnel in Monrovia.[17] The Bureau's work culminated in a 1952 reconnaissance study of the Mt. Coffee dam project.[18] Actual construction of the dam, carried out by the Raymond-Utah Company (a subsidiary of the engineering firm Utah Construction and Mining), was characterized by delays, political struggles, and violent labor unrest. In addition to a project manager who made "flagrantly undiplomatic statements" about the 1,500 Liberian workers on the dam, Raymond-Utah was clearly squeezing its workforce by, for example, refusing to adequately pay for overtime and failing to provide transportation for manual laborers. Liberian newspapers noted the "appalling working conditions" at the site and the "racist management" who were apparently contemplating "reprisal against leaders of strike."[19] On October 27, 1965, over 1,000 Liberian workers went on strike at the dam site due to the company's failure to provide overtime pay. Despite these hurdles, the dam was completed in late 1966. Plans to triple the project's 30-megawatt generating capacity were wiped out in 1990 when the dam was taken over and damaged by rebel forces during a civil uprising. Most recently, the Liberian government received a US$65 million loan in January 2013 from the European Investment Bank to repair the dam and expand its hydropower capacity.[20]

Nigeria

Outside of Ethiopia and the Blue Nile investigations, the Bureau of Reclamation's engagement in Nigeria's Lake Chad basin constituted its most intensive program in Africa. For three years, from 1965 to 1968, a nine-person Bureau team undertook a "reconnaissance investigation of land and water resources" in the vicinity of Lake Chad in the country's semiarid northeast. The study identified four potential irrigation projects that would cover an area of 140,000 acres and were expected to confer several benefits, including reduced flooding, expanded agricultural production, crop diversification, and enhanced employment opportunities.[21] The Lake Chad program generated little geopolitical interest on the part of the US government. However, a contemporaneous Nigerian project—the Kainji Dam on the Niger River roughly 1,000 km upstream from its mouth—became the focus of State Department discussions beginning in 1963, when an American firm (Standard Overseas International) unsuccessfully bid on the dam's construction. Confirming the geopolitical and political-economic goals of US support for water resource development, an embassy official stated at the time that the United States is "anxious, for political as well as commercial reasons, to see the main contract for the dam awarded to an American contractor." To this end, the State Department advised the Department of Commerce that any information coming out of the American embassy in Lagos deemed "unclassified" could be passed along to American companies interested in bidding on the dam's main contract.[22] Ultimately, the dam was built without the participation of any American firms.

Somalia

Although it was a very limited engagement, the Bureau did undertake a brief review of water resources development for the government of Somalia (funded by USAID) in 1963. The subsequent report stressed the need for improvements in administration of the relatively modest national water resource development program, which consisted primarily of the enhancement of potable water supplies through well drilling in rural areas. The US government, largely through the US embassy, had offered financial and technical assistance to the Somali government since 1954.[23]

Tanzania

In an effort similar to the Kano Plain reconnaissance survey in Kenya, a three-person Bureau team visited Tanzania in late 1966 to assist in the review of "existing studies of the Rufiji Basin, in carrying out new predevelopment studies, and in providing guidelines relating to the formulation of the Rufiji Basin Development Authority." In the words of the Bureau experts, the Rufiji River basin, the largest in Tanzania, covering about 20 percent of the nation's territory, "unquestionably has the resources to support major self-liquidating programs in the future," including major hydroelectric dams and irrigation projects.[24] In addition to summarizing and expanding the government's plans for development of the basin, the Bureau team suggested a number of administrative and legal reforms regarding water resource planning and development and recommended that a basin authority be created along the lines of the TVA. The mission also identified the Stiegler's Gorge Dam as a priority project, which prompted a long national debate in Tanzania about water development in general.[25]

Asia

Afghanistan

The participation of the United States in the Helmand River project in Afghanistan remains a pertinent example of the commingling of technical assistance, geopolitics, and water resource development during the height of the Cold War.[26] The study, planning, and construction activities in the Helmand Valley in the 1950s and 1960s were coordinated by economic development personnel affiliated with the US Point Four program and Morrison Knudsen, which at the time was America's largest civil engineering firm. Hired by the Afghan government in the late 1940s to develop the Helmand River's hydroelectric and irrigation potential, the *Emkayans* (as they styled themselves) promised to bring "another backward land into the 20th century."[27] Overall, the United States expended over US$100 million on development initiatives in the Helmand Valley from 1957 to 1979.[28] Other than the consulting assignments carried out by John Savage in the late 1930s and early 1940s, the Bureau's role in Afghanistan was severely limited, a situation that galled Floyd Dominy, who, despite the photograph of himself and an Afghan engineer on the crest of the Arghandab Dam in 1963, criticized the State Department's lack of consultation with the Bureau on the Helmand and other international aid programs. The "foreign affairs" people, according to Dominy, came to the Bureau for assistance on remedying the "mess" in Afghanistan in the late 1950s only "because the Afghan government told [Point Four officials] that perhaps the Bureau of Reclamation ought to be consulted" since it was the "one agency in the Federal government that knows how to do these things."[29]

Bangladesh/East Pakistan

While not the primary agent in this instance, the Bureau nevertheless played a catalyzing and oversight role in what would become a prominent and ultimately controversial large dam, the Karnafuli project on the river of the same name in southeastern East Pakistan (now Bangladesh). Survey work on the Karnafuli project began in the early 1950s. A US-based firm, the International Engineering Company (IEC), was retained by the US Foreign Operations Administration (FOA), a short-lived US technical assistance agency, and the government of Pakistan to undertake a study of electrical development in East Pakistan, which was completed in December 1954. This was part of a larger study carried out by the IBRD (World Bank), which also included an economic investigation of the Karnafuli hydroelectric dam.[30] Almost immediately after these studies, the East Pakistan government began pressuring both the United States and the IBRD for funding to proceed with construction of the project. In the words of a US embassy official, Pakistan was "pressing" to use the funds allotted to it by the FOA to pursue the construction of the Karnafuli dam. The embassy's country director (under the Point Four program) proposed that US$4 million be made available to extend the contract of IEC, "who would enter into immediate negotiation with few selected US contractors for construction [of the dam] . . . with IEC to provide detailed engineering services and supervision during total construction period."[31] The country director pointed out that either the FOA or the Bureau of Reclamation would facilitate negotiations with contractors and advise the government of Pakistan on selecting the most appropriate bid.

Given the preliminary design work made available by IEC's initial studies, Pakistani officials forged ahead with the project, a decision criticized by the American contractors. IEC chief engineer T. Mundal bemoaned the fact that the government of Pakistan was "proceeding with the construction of the Karnafuli project based on the study drawings of our Reconnaissance Report" when these designs had "advanced considerably" since that report and "design of the structures has changed substantially."[32] Washington communicated to East Pakistan in no uncertain terms that this situation was unacceptable. Additionally, according to James Baird, acting director of USOM, the aforementioned recommendation to Washington to provide FOA funds for proceeding with the Karnafuli project was "conditioned on the willingness of the government of Pakistan to have the actual construction of the project performed by a United States contracting firm." While the government of Pakistan apparently acceded to US calls for patience, by March 1955 Said Hasan, the Pakistani minster of economic affairs, expressed dismay at the delays in cost estimates for the project from IEC. The delays in the Karnafuli project, he said, were causing "a great deal of anxiety to Government." Karnafuli, argued Hasan, "has come to occupy a unique position in the minds of the people of East Pakistan and is at present completely identified with their concept of economic advancement."[33] Throughout these politically charged negotiations, the Bureau provided consultations on the project, although not in the form of a direct presence by a Bureau mission. Ultimately, the Kaptai Dam on the Karnafuli River was completed in 1969. It displaced over 100,000 people, many of them ethnic minorities who were never compensated and subsequently faced discrimination and reduced livelihood opportunities in official resettlement areas, some located across the Indian border.[34]

India

The majority of the Bureau's involvement in India coincided with the early days of development of the Indus River valley in the 1950s.[35] Bureau consultants played a moderately important role in advising the Indian government's engineering experts on the Bhakra Dam on the Sutlej River and the Beas Dam on the Beas River, both part of the Indus River basin shared with Pakistan. As early as 1945 John Savage was requesting support from the Department of the Interior for the Bureau to become directly engaged in designing the Bhakra Dam at the request of the Indian government.[36] From 1962 to 1965 the Bureau carried out several studies on the feasibility and technical design of the multipurpose Beas Dam in the Punjab at the request of the Indian government. These studies focused on the suitability of local geological conditions and the source materials that would be used to form the dam as well as on the economic feasibility of the project.[37]

Korea (South)

The Bureau became directly involved in Korean projects around 1950, when it created a team called the Korean Hydroelectric Power Feasibility Survey Group following an agreement with the Economic Cooperation Administration.[38] This team examined several potential sites for hydroelectric development, and its subsequent report encompassed, according to engineer Robert Herdman, "a review of the Japanese data that has been found, an appraisal of its usefulness, a determination of the relative merits of each project, the course of procedure required to accomplish a second stage in the investigation, the American personnel required, and the estimated cost of preparing designs, cost estimates, and project report for each project."[39] By the mid-1950s, following the Korean War, in which most major power plants in the country, including several hydroelectric facilities in the north, were destroyed, South Korean leaders confronted an energy crisis. The generation of electricity was seen as a crucial aspect of rebuilding the Korean economy and society after the conflict, but reconstruction of the power infrastructure faced significant management and organizational hurdles.[40] In short, Korea appeared ripe for the delivery of American expertise and technical advice.

By the late 1950s, the Han River basin, situated in the northern third of the country, became the focus of intense interest by the Korean state due to its immense hydroelectric potential as well as possibilities for multipurpose development. By 1961 Korea had developed only 10 percent of its hydroelectric potential, and the Han River basin encompassed the "most suitable sites."[41] The government envisioned a chain of power plants on at least two of the Han system's main tributaries (the North Han and the South Han). Two hydroelectric dams had already been developed on the North Han, and four sites on the South Han were "under active consideration." Under the sponsorship of USAID, the Bureau carried out an intensive study of the Han River basin from 1966 to 1971, during which nine of its experts engaged in a water resources development survey. One of the primary goals of the Bureau team was to train its Korean counterparts (about 75 Koreans were involved) in all techniques of "water resources investigations and planning."[42] A preliminary evaluation of the proposed survey pointed out the paucity of data on the Han River and criticized the "piecemeal plans" of the Korean government, which had "not been made with a view toward coordinated, multipurpose basin development."[43] The Han study was deemed a success by 1975, by which time "the Korean agency now has the capability of conducting similar studies without relying on technical assistance from outside organizations."[44] The development of the Han River went ahead as planned, and its flow is now controlled by six major hydropower installations, three multipurpose projects, and one flood control dam, nearly all of which were proposed in the Bureau's 1971 report.[45]

Laos

Aside from its obvious involvement in the Bureau's Mekong-related development planning, most notably as the site of the proposed Pa Mong Dam (see chap. 5), Laos was the focal point of technical efforts to spearhead Mekong development that were deeply tinged with US geopolitical calculations. Beginning in roughly 1954 and continuing throughout the 1960s, the US government perceived Laos as key to the long-term stability of Southeast Asia, in part due to its geography: both the Eisenhower and Kennedy administrations saw Laos as a buffer between looming Communist China and the other states of mainland Southeast Asia. Yet foreign policy officials badly misjudged the conditions within Laos that would stymie US efforts to prop up and maintain its anti-communist regimes. Eventually, in tandem with the US failure in Vietnam, the Soviet- and North Vietnamese-backed Pathet Lao assumed full control over the country in 1975.[46]

The circumstances surrounding the financing and construction of the Nam Ngum Dam—constructed in the early 1970s during the height of the Lao civil war and to this day the country's single largest power-producing operation—exemplify the geopolitical thinking that guided even a rather modest hydroelectric project. The Nam Ngum dam—identified in a 1959 Mekong Committee–sponsored survey as a top-priority tributary project in the Mekong basin—is a concrete gravity-arch structure with a height of 75 meters. Over 80 percent of the 150 megawatts of electricity it generates is sold to the Electricity Generating Authority of Thailand; the remainder is dedicated to powering the Lao capital, Vientiane.[47] As the conflict in Vietnam intensified in the mid-1960s, concerns over the incursions of the Pathet Lao into the Mekong Valley region prompted the US government to speed up plans to offer development assistance to its beleaguered proxy government in Vientiane. Lyndon Johnson singled out the Nam Ngum project in his Mekong River basin speech of 1965 (see chap. 5), and negotiations carried out by United Nations and Mekong Committee representatives resulted in a 1965 agreement among Laos, Thailand, and an array of donor countries to fund, design, and build the Nam Ngum dam. The United States pledged just over US$12 million to the project, over 50 percent of its total cost.[48]

Throughout its conceptualization, financing, and construction, the Nam Ngum project confronted significant technical and political obstacles. After the signing of the agreement in 1965, the World Bank stepped in to administer the funds, but the bank regarded the project as "uneconomic."[49] One observer noted in 1968 that the Americans had "not been particularly pleased" at the prospect of providing 50 percent of the financing.[50] At some point during the dam's feasibility studies, special American survey parties—presumably from the Bureau team already in the region—were called in to ensure the project's general feasibility. In addition, State Department officials skeptical of the Johnson administration's seemingly uncritical support for the dam pointed out the lack of a market for electricity in the area of the dam (about 100 km northwest of Vientiane) and the potential security risks associated with a high-profile international project being constructed in a war zone. Others countered that the Pathet Lao and the Vietcong had "nothing to gain by the destruction of a valuable nonstrategic installation which may sooner or later pass into their own custody," and argued that "once power and water are made available, they have a way of creating their own demand, and in that very process of working desirable social and economic change."[51] Indeed, throughout the phase of construction, from 1966 to 1971, in a region controlled by the Pathet Lao, the insurgents left the Japanese and Canadian workers building the dam largely unharmed.[52] Although the Bureau's involvement in Nam Ngum was negligible, the project nevertheless figured prominently in the broader US geopolitical designs for the Mekong region.

Pakistan/West Pakistan

Pakistan has served as a crucial node in both the past and present geopolitics of large dams, owing in part to its arid climate and long history of water resource development. In the portion of the Indus River basin lying within Pakistan (another part of the basin lies within Indian territory), coordination among the State Department, the World Bank, and several US-based engineering firms played key roles in river basin planning and dam construction throughout the late 1950s and into the 1970s.[53] While the Bureau of Reclamation's official involvement may have been negligible, Bureau engineers serving in a consulting capacity with the World Bank played prominent roles in the Indus water development program,[54] particularly in the 1950s leading up to the international water-sharing agreement between India and Pakistan penned in 1960.[55]

By the early 1960s, Pakistan's desire for US financial and technical assistance for completing the Tarbela Dam, a large hydroelectric project to be located on a tributary of the Indus, led presidential adviser Robert W. Kennedy to argue that Pakistani officials "have come back strong at us and other Indus clubbers on Tarbela Dam," and that the United States has "so many problems with [the Pakistanis] that holding out Tarbela as a possible consolation prize may serve our political interests."[56] These "many problems" and "political interests" were outlined further in several internal White House documents concerning State Department Under Secretary George Ball's mission to visit Pakistani president Ayub Khan in September 1963. Ball was advised, in essence, to use financing of the Tarbela Dam (packaged with military hardware and other assistance) as a bargaining chip in an effort to convince the Ayub regime to reverse growing Chinese influence in Pakistan, which the US State Department interpreted as a move by the Pakistani state to strengthen its position in the ongoing conflict with India over Kashmir.[57] The Tarbela Dam was eventually financed in 1968 through the World Bank and other donors with US support and was completed in 1974.[58]

Throughout the same period as negotiations over the Tarbela project, the Mangla Dam also became a source of conflict that revolved around Pakistan's desire for highly visible, expensive water development projects that would ostensibly meet its power and irrigation needs and the United States' wish to draw a politically volatile postcolonial nation into its sphere of influence. As early as 1955 US embassy officials in Karachi lobbied the State Department to fund an engineering study by the Bureau of Reclamation focused on a dam site on the Jhelum River. The embassy dutifully noted Pakistani government claims that the proposed project, the Mangla Dam, was "not repeat not involved Indus Basin dispute, but in accord recommendations World Bank."[59] US perceptions of Pakistan as a key regional actor in larger Cold War struggles over Central and South Asia virtually demanded that State Department officials and White House advisers take a special interest in seemingly inconsequential development projects, albeit ones with significant costs and impacts. With funding from the World Bank and the Asian Development Bank, the Mangla Dam was constructed between 1961 and 1967 with the participation of a consortium of eight major US engineering firms. The dam, now the ninth largest in the world, led to the displacement of approximately 90,000 people.[60] Both the Tarbela and Mangla projects have drawn significant criticism for their socioecological impacts.

Philippines

Given the Philippines' long history of colonial, postcolonial, and neocolonial relations and numerous instances of political, educational, and cultural exchange, it is not surprising that the Bureau found development officials and engineers in that country (many of whom trained in the United States) who advocated the rapid exploitation of its rivers via multipurpose projects and basin-wide development and who desired US assistance in bringing these plans to fruition. One of the earliest American interventions in water resource development in the Philippines fell outside the ambit of the Bureau. In 1953 American engineers working with the California-based Guy F. Atkinson Company began work on the massive Ambuklao Dam on the Agno River in the Cordillera region on the island of Luzon. The Ambuklao project, the "largest hydroelectric development project so far undertaken in the Philippines," was financed in part through a US Export-Import Bank loan. The project was also thoroughly American: dam design was carried out by the Harza Engineering Company of Chicago (in an advisory role to the Philippine government's National Power Corporation), the power generators were furnished by General Electric, and national plans to develop the country's hydroelectric potential were hatched in the late 1940s by Westinghouse International Electric Company's survey of the islands.[61] The dam, completed in the early 1960s, was an important precursor to the Bureau's efforts in the 1960s and 1970s to undertake comprehensive river basin development plans for the Philippines' major river systems.

The Bureau carried out one of its more sustained engagements from 1963 to 1967 when it advised the Philippine government on river basin development. The outcome was a massive study of the development potential of seven river basins: the Cagayan, Pampanga, Bicol, and Agno Rivers on the island of Luzon, the Ilog-Hilabangan watershed on Negros, and the Agusan and Cotabato Rivers on Mindanao. As proposed in a series of reports, development of the basins was designed to "provide flood control, water supply for irrigation and domestic use, hydroelectric power generation, and other related functions."[62] A 1966 cover letter from Commissioner of Reclamation Floyd Dominy described the promising development potential of the Central Luzon basin and observed that the basin was "just now entering that point of time in history when large scale programs must be undertaken or the social and economic welfare of the areas will not progress in line with the needs of its people or the nation."[63] River basin development and economic development thereby went hand in hand. As was the case with nearly every Bureau operation, the public record of achievements within the rubric of the seven basin studies belied the behind-the-scenes assessments of US officials. A progress report in January 1964 by the head Bureau engineer, D. R. Burnett, outlining the Bureau's work spoke of the "knowledgeable individuals and organizations in the Philippines" who were "eager to assist in every way possible," and asserted that there had been "no overtones of jurisdictional arguments."[64] Yet an evaluation of the progress of the Bureau investigations undertaken by a USAID official eighteen months later was less optimistic. Noting that the Bureau had "devised an excellent project plan for the Upper Pampanga River" in Central Luzon, the official directed attention to the numerous "obstacles and impediments" that were "institutional and political in nature." The planning, implementation, and administration of water resource development in the Philippines was "defused through a welter of independent and largely uncoordinated government bureaus, agencies and authorities." More problematically, he reported, "water resource development programs have become so intimately associated with partisan politics" that sound programs of water development were "near impossible," and the government's position on the "reimbursability or non-reimbursability of specific project purposes" was problematic.[65] Despite these and other problems, the official recommended that USAID continue to support the Bureau's work in the name of maintaining strong US-Philippine relations. Geopolitical expediency again assumed precedence over the dynamics of local conditions that in the end would undoubtedly compromise the goals of river basin development.

By 1975 the irrigation components of the seven basin schemes were "in varying stages of development,"[66] but the Bureau's grand plans for the seven basins never materialized at the scale laid out in its reports for financial, economic, and political reasons. Those multipurpose projects that were completed—including the Magat and Agusan dams—have generated a myriad of socioecological problems, have largely underperformed, and in some cases have generated social conflict—as have other projects that were never completed. For example, four large hydroelectric projects proposed for the Chico River, a prominent tributary of the Cagayan River in the Cordillera region of northern Luzon, were originally identified in the Bureau's report on the Cagayan basin. The plans for transforming the basin are on their own merits astounding: the Bureau team identified 38 sites with potential for large-dam construction, with five proposed dams measuring over 200 meters in height. The report on the Cagayan basin foresaw "opportunities for multiple power developments" via dams and reservoirs and even, in some instances, "through tunnel diversions." The dam sites on the Chico River were perceived as the best sites for future development.[67] When the economic and political prospects for damming the Chico River improved in the early 1970s, the dictatorial regime of Ferdinand Marcos secured a World Bank loan for preliminary project funding to build four massive hydroelectric dams (with a total installed capacity of 1,010 megawatts) and one more modest irrigation structure. The proposals quickly engendered opposition from the various ethnic minorities of the region drained by the Chico River, primarily peoples of the Bontoc and Kalinga groups, who would be displaced by the projects and whose ancestral and culturally significant lands would be inundated. In a series of protests and actions that turned increasingly violent as government survey teams, bolstered by military security forces, undertook preparatory work for the dams, the affected peoples of the region formed intertribal coalitions and allied themselves with the New People's Army, the armed wing of the Communist Party of the Philippines, which had been carrying out a general insurgent action against the Marcos regime since 1969. Amid intimidation campaigns and military abuses, the Kalinga and Bontoc opponents of the dams also attacked government personnel, and by 1980 the Chico Valley "had become a virtual war zone."[68] Eventually the Marcos regime effectively canceled plans for the dams, which was perceived by many as a victory for the then-nascent global anti-dam movement. The ambitious river basin development scheme for the Cagayan basin fleshed out in the mid-1960s by Bureau experts produced little more than violent struggle.

Sri Lanka/Ceylon

One of the earliest Bureau efforts to disseminate river basin planning took place in Ceylon (now Sri Lanka) almost simultaneously with the expansion of US technical assistance under Truman's Point Four program. As one of the first Bureau forays into international development, albeit one in which Bureau involvement was rather limited, it confirms the technopolitical character of these exploits. Before the formalization of the Bureau's international efforts in 1950, a Bureau engineer, Paul von der Lippe, served as an adviser to the Ceylonese Ministry of Agriculture and Lands in the late 1940s. During this same period John Savage and John Cotton advised the government on flood control measures as consultants. In addition, the Gal Oya Dam, the initial development project on the Gal Oya River, was overseen by Morrison Knudsen International. Morrison Knudsen received the US$15 million contract to build the Gal Oya Dam in early 1949, and its staff of 60 American supervisors and 1,400 Ceylonese laborers started construction on the project in April of that year.[69] As was the case in many of the dam construction sites in the tricontinental world, Morrison Knudsen personnel built their own "typical United States hamlet for themselves and their families" in the jungle near the dam and reservoir location.[70] The Gal Oya undertaking was inaugurated in 1949 with the following words from Prime Minister D. S. Senanayake: "Gal Oya has become almost a household word. It is symbolic of New Lanka. May it obtain fulfillment speedily and herald the progress of our march towards self-sufficiency."[71] The government's vision of Gal Oya development concentrated on irrigating Ceylon's vast Dry Zone lands, regions in the eastern and northern parts of the island that were the site of Ceylon's great irrigation works two thousand years previously. In 1954 geographer Clifford MacFadden spoke in glowing terms of "Ceylon's little TVA" in the Gal Oya valley. MacFadden himself served as chair of geography at the University of Ceylon, a position that was jointly sponsored by the State Department under the Smith-Mundt Act and the University of Ceylon.[72]

Taiwan

Following the collapse of the Chinese Nationalist government in the late 1940s and its exile to the island of Formosa, the Bureau maintained its technical assistance to the newly formed nation of Taiwan. Bureau engineers guided many of Taiwan's early forays into developing its territory's latent hydroelectric potential, and electricity produced by dams was a key element of the new state's efforts at reconstruction and industrial development in the two decades following World War II. A prime example is the Wu-Sheh Dam, located on the river of the same name in central Taiwan. The project was initiated by the Japanese colonial regime—which had installed a power facility and begun excavating the site—in 1944 during occupation, and the Eisenhower administration's Mutual Security Agency requested Bureau assistance to complete the project in 1953. The six-person Bureau team in Taiwan provided comprehensive technical assistance, including feasibility studies, dam design, equipment and construction recommendations, cost estimates, and training of "Chinese" personnel. The 115-meter, curved gravity concrete structure was complete by 1958 in spite of several environmental challenges involving landslides and seismic activity.[73] The Wu-Sheh scheme established an enduring partnership between Bureau engineers and their Taiwanese counterparts that continued into the 1960s and 1970s. It was also a direct outcome of US geopolitical designs for Communist China, as was made clear in Far East security policies of the time. The Mao regime saw economic relations and a presumed security pact between the US and Taiwanese governments as direct threats to its security interests, and its concern was stoked by US Secretary of State John Foster Dulles's constant refrain of "unswerving" support for the Chinese Nationalists in the early to mid-1950s. The "Taiwan issue" precipitated a series of geopolitical crises in the region during this period.[74] State Department officials perceived Bureau activities, along with other economic and technical assistance programs, as important components of basic US foreign policy objectives in Taiwan, which included "to deny Taiwan . . . to the Communists," to promote a "a friendly, stable, responsible Free Chinese Government enjoying the broadest possible base of public support," and to bolster Taiwan's military capabilities.[75]

Accordingly, the Bureau participated in significant preparatory work for two additional hydroelectric dams, the Shihmen project in the mid-1950s and the Tsengwen Reservoir project in the 1960s. Throughout this period Taiwanese officials pressed the United States vociferously for financial support for water resource development—for the projects mentioned above and another major hydroelectric scheme, the Tachien project—through the ICA and eventually USAID. The ICA eventually contributed $US4 million to the Shihmen project, but not before requesting the Bureau to undertake a comprehensive "review of the preliminary project reports, plans, and data"—not, in the words of the ICA, to seek "justification of support for the project," but to gain a "fair evaluation of benefits."[76] Again, the Bureau was being asked to provide expert verification of a project to be funded to some degree for geopolitical purposes. Its report concluded that, despite the need for some follow-up studies of the power market, the project was "well planned, economically justifiable, and financially feasible."[77] The Bureau offered a similar consultancy for the Tsengwen multipurpose project, constructed on the Tsengwen River in west-central Taiwan in the mid-1960s.[78] The final salient Bureau contribution to "Free China" came in the form of an expert review of the "nation's water resource development and management policies" in 1965 at the bequest of USAID and amid growing awareness that Taiwan was facing several interrelated water challenges (e.g., shortages and competing demands from different users). In line with its own methodologies directed toward domestic water issues, the Bureau report suggested a more "regional" approach to water development, bolstering the skills of personnel at "intermediate or middle management" levels, and greater coordination of activities across "national, provincial, and local spheres."[79] As in many countries throughout the world, the Bureau advocated reorganizing a water bureaucracy largely in its own image.

Thailand

The Bureau of Reclamation's relationship with Thai engineers and government officials spans decades and represents one of the most enduring international engagements of the Cold War era of technical assistance. Beyond the Bureau's Mekong basin–related activities (see chap. 5), the Bureau provided design and construction expertise to the Thai government on Southeast Asia's first massive hydroelectric dam, offered administrative advice to Thailand's water bureaucracies, and devised an ambitious river basin development scheme for the nation's northeastern region over a period running from the early 1950s to the early 1970s. In no small part because of Thailand's geopolitical positioning, US technical and economic assistance programs were intimately linked to global and regional concerns over the containment of communism. A blunt statement by the director of USOM in Thailand in 1956 captures the essence of US economic assistance in the country: its goal was "to help the Government of Thailand in furthering economic—and hence political stability. The US aid program is also intended to strengthen Thailand's position as a center of noncommunist influence in the Asian area."[80] Military, economic, and technical assistance from the United States, oftentimes funneled through the World Bank, was thus crucial to Thailand's economic development and to American geostrategic interests from the 1960s to the 1980s.[81] The Bureau played a crucial part in sustaining Thai-US relations, and the earliest manifestation of this role revolved around technical support for the Yanhee multipurpose dam project.

In the early 1950s Thailand's Royal Irrigation Department, the country's primary water resource development agency for the first half of the twentieth century, requested that the Bureau review the studies that it had undertaken regarding the proposed Yanhee Dam (later renamed Bhumiphol, after the Thai monarch), a massive multipurpose scheme targeted for the Ping River in the province of Tak. Although the Thai government envisioned some irrigation, flood control, and navigation benefits, the primary aim of the dam was hydropower "for use in the main population centers of Bangkok and Thonburi" and other urban concentrations in central and northern Thailand.[82] The Bureau evaluation—carried out as a condition of the US$66 million IBRD (World Bank) loan requested by the Thai government—concluded that the "engineering feasibility" of the dam was "good" and that the 154-meter concrete arch dam, with an installed capacity of 750 megawatts, would indeed be "economically superior" to other forms of electricity production and other "hydroelectric developments." While no such recommendation appeared in the official evaluation, the investigation by the seven-person Bureau team in 1953 resulted in abandonment of the dam site originally proposed by the Thai government in favor of a more geologically and topographically advantageous location.[83] Throughout the period between publication of the Bureau's report in 1955 and commencement of construction of the Yanhee Dam in 1958, the project was singled out in US embassy and State Department communiqués as one of the prime examples of how economic and technical assistance could cement constructive US-Thai geopolitical relations within a volatile region. For example, as the loan to Thailand was under consideration by the World Bank in mid-1956, a State Department official in charge of Southeast Asian affairs entreated his superiors to "mention to . . . the IBRD the desirability of financing" the dam, and stated that "if the IBRD could approve this project in the near future it would mean a great deal to Thailand and to our objectives there."[84] Later that year the State Department made clear to the US embassy in Bangkok that it was "aware the [Thai] Prime Minister considers this project [Yanhee] high priority."[85] Eventually the World Bank approved the loan, and the dam was completed in 1964, following the displacement and resettlement of over 24,000 people.[86]

Apart from the Yanhee project (and outside the rubric of its Mekong work), the Bureau's activity in Thailand was focused on an ambitious study of Northeast Thailand's two major river basins. The northeast had been targeted as an economically and socially "backward" region susceptible to communist infiltration from nearby Laos and Cambodia. An eight-person Bureau team commenced work on a reconnaissance survey of the Mun and Chi River basins in late 1964. The primary aims of the studies were to "recommend a program for the orderly development of the basins" and to prioritize potential dam sites according to their feasibility.[87] Two aspects of the Mun and Chi basin studies, one linked to the geopolitical ambitions of the American state in Southeast Asia at the time and the other to the influence the studies exercised over the Thai state's future plans for regional development, underscore their technopolitical significance. Thai officials were already aware by the mid-1960s that most of the large-scale water development projects initiated as part of the "Northeast Development Strategy" brought few benefits to rural people. On the advice of US officials, a series of community-focused development activities were launched in 1968 under the auspices of key Thai ministers and cabinet members, and an advisory group was set up under the technical assistance program of the US government. A good deal of this assistance involved small-scale water projects, and US involvement was clearly aimed at increasing the resources available in Northeast Thailand, based on the assumption that this would dampen the influence of communism in the region.[88] The Mun and Chi basin investigations must be seen as part of this broader "development as counterinsurgency" strategy. In the longer term, the Mun-Chi studies set in motion a decades-long effort by the Thai government to transform the entire northeastern region through irrigation and power development in the service of an ambitious vision of agro-industrialization. In fact, Thailand's participation in the Mekong project and the plans surrounding the Pa Mong Dam (see chap. 5) were predicated in large part on the capacity of Mekong development to stimulate development in Thailand's portion of the basin. The Bureau studies in Thailand in the 1960s established the blueprint for the numerous water development schemes carried out by the Thai government in the decades that followed, largely to the social and ecological detriment of the peoples and river systems of the northeast.[89]

Middle East

Iraq

During his trip to Iran in 1952 (see under "Iran" below), Anthony Perry also visited Iraq, where he briefly discussed river basin development with some Bureau engineers assigned to the Point Four program in that country and with Iraqi water resource specialists. His conversation with Tariq Al Askari, a member of Iraq's Development Board, familiarized him with three irrigation projects being contemplated in Iraq, one of which was being designed by the Harza Corporation of Chicago.[90] What eventually became the Derbandikhan Dam, on the Diyala River some 230 kilometers northeast of Baghdad, was beset by technical and financial difficulties during its construction in the late 1950s, and the construction contract with Harza was terminated in 1959 due to the Iranian government's dissatisfaction with progress on the dam.[91] Perry's visit ushered in a series of additional consulting visits by Bureau engineers to Iraq in the 1950s. For example, U. V. Engstrom served for ninety days as the official Bureau of Reclamation "liaison" to Iran's politically influential Development Board in 1954. Engstrom's reports to the Bureau's Foreign Activities Office in Denver offer a rare glimpse of some of the daily frustrations of the individuals involved in overseas technical assistance. He suggests that pre-trip briefing could be much more comprehensive and obviate the "anxiety" experienced "until streets are known and ways of getting around [in Baghdad] are established."[92] He also notes that nearly every employee of Iran's Development Board engaged in water resource development is a former employee of an (often American) consulting firm, identifying yet another link between the geopolitics and political economy of water-related technical assistance.

Iran

Efforts to encourage river basin development in Iran throughout the 1950s and 1960s were connected to US-Iran geopolitical relations and illustrate, yet again, the deeply politicized character of river basin planning—and more broadly, technical assistance—and its use as a strategic tool during the Cold War. Early involvement of the United States in water resource development in Iran was focused on the Karaj Dam, a massive hydroelectric scheme conceived and designed with technical assistance provided by the Bureau of Reclamation in the early 1950s. Karaj was one of the United States' "model" Point Four projects, and it occupied an important symbolic position within Iran as a critical project of modernization. Following requests to the State Department for technical assistance from the Iranian government, Bureau engineer Anthony Perry traveled to the Middle East in 1952 and commenced a survey of Iran's hydroelectric potential. Perry examined the feasibility of projects in "water supply, irrigation, industry, civil aviation and communications." While at the time work on the Karaj Dam—carried out by a French construction firm—had ceased, Perry observed that the Karaj project "would have a tremendous economic impact on the City of Tehran and the surrounding communities," and that "if funds could be made available to Iran for public works, this project should be made the first one to be undertaken."[93] Although State Department officials expressed reservations about the benefits of dam-related financial assistance in strategic terms, a later Bureau report expressed a strong preference for proceeding with the project, stating that "development of water resources through storage on the Karaj River is essential to the economy of Tehran and the Karaj River area."[94]

This skepticism over the Karaj project belied American geopolitical interests in Iran and the broader region. Iran figured prominently in post–World War II foreign policy, not least because of its prodigious oil supplies. US economic assistance was predicated on the assumption that a pro-Western regime in Tehran would be beneficial in terms of securing critical energy resources.[95] As noted in a report commissioned by the National Security Council in 1954, the United States had "provided $70 million of emergency economic aid and approximately $23 million of limited technical and economic assistance aid" since the assumption of power of the Zahedi regime.[96] The report also set aid and technical assistance to Iran in a regional context, pointing out that for "practical and psychological reasons, we . . . should fully recognize that if Iran gets the impression that it is our concept that Turkey, Pakistan and Iraq, strengthened by United States aid, are to be counted upon by us to defend Iran, the deep resentment which would result would do irreparable harm to the attainment of United States objectives in Iran."[97] Ultimately, the State Department leveraged approval of a US$30 million loan from the US Export-Import Bank, and the Karaj project was constructed by the American engineering firm Morrison Knudsen after a great deal of political haggling involving Iran's primary development agency, Plan Organization, over whether or not a US firm would indeed be awarded the contract.[98]

Aside from the Karaj project, essentially a single large scheme to provide electricity and water supply to Tehran, the scheme in Iran that most dominated US thinking about dams and river basin planning was an ambitious program of dam and irrigation infrastructure to be implemented in the Khuzestan region under the guidance of David Lilienthal's Development Resource Corporation (DRC) in the 1960s.[99] Iran's Plan Organization secured a US$42 million loan from the IBRD in 1960 to launch the initial program, which included the construction of the Mohammed Reza Shah Pahlavi Dam (since renamed the Dez Dam) on the Dez River, the erection of a high-voltage transmission system connecting the Dez power plants with five cities in Khuzestan, the creation of a sugarcane production complex consisting of 10,000 hectares of irrigated farmlands, development of an additional 20,000-hectare pilot irrigation project as a precursor to the much larger Greater Dez irrigation project, and several minor projects.[100] Despite the view of both Lilienthal and the Iranian state that development of the basin would help modernize the entire country, the Khuzestan basin development process was bedeviled by mutual mistrust between the Iranian government and the DRC consultants, conflicts over the awarding of dam contracts to British and American firms, and interagency struggles over water resource governance within Iran. Although several dams were constructed under the rubric of the Khuzestan basin scheme—including the aforementioned Dez Dam—the comprehensive basin development envisioned by Lilienthal and the Iranian government never materialized.

Jordan

The Bureau published a reconnaissance study on the potential for undertaking comprehensive river basin development of the Yarmouk-Jordan Valley in 1953,[101] but the plan languished for over a decade due to regional geopolitical constraints and lack of government funding. As it was being implemented in the late 1960s, the Jordanian agencies charged with river basin development encountered several snags. The government was interested in building the Khalid Bin al Walid Dam on the Yarmouk River, but locating financing for construction of the project proved difficult. While there was some level of commitment from Arab League states, it was contingent on additional funding from different sources. US embassy officials in Amman, wary of the Aswan case and hopeful that American firms might enter the bidding process, reported that it was "virtually certain that Soviet or Czech firms" would "submit a politically-motivated low bid."[102] The crux of the issue for the United States was whether US funding was necessary to achieve political objectives. As noted by an embassy official in 1967, the Jordanian government "could . . . approach the USG [US government] for financing," given that the "project is sound and entirely consistent with our development objectives in Jordan, [and of] crucial importance to the optimum usage of [present irrigation schemes], as well as further irrigation development in the Jordan Valley." The official questioned, however, whether the United States could indeed finance the dam given that contracts for other components of the project had "already been awarded to non-US firms." Despite concerns that the "Soviets would be receptive" to an offer to finance and build the dam, the embassy official felt it was likely that the Jordanian government "would be seriously inhibited from choosing this alternative" because of its "apprehension that Soviet funding could affect US aid toward Jordan."[103] The Jordan example is representative of the geopolitical algorithms frequently debated within the State Department over funding for large dams, irrespective of technical assistance.

Turkey

One of the Bureau's longest-lasting and, in its terms, most successful international programs involved an advisory team of water resource experts that consulted with the Turkish government on a wide range of issues from the early 1950s through the mid-1960s. Turkey's relationship with the Bureau started modestly in the early 1950s when Turkish engineers, sponsored by the Mutual Security Agency, underwent training at the Bureau offices in Denver. After this initial encounter, the Bureau became deeply involved in the evolution of water resource development in Turkey. To cite one example, Turkey's primary water development organization (Devlet Su Isleri, or DSI) was modeled directly on the Bureau of Reclamation. Moreover, the institutional setting created for water resource development in Turkey, including national legislation, was based in part on the advice of Bureau experts in the 1950s. By 1955 the Bureau advisory team had grown to ten technicians, who were subsequently reassigned to the ICA and renamed the River Valley Development Team, reflecting the emphasis of the mission members on translating the techniques of river basin planning and development into a Turkish context.[104] The contact between Bureau personnel and DSI engineers was prolonged and familiar; in the words of one report, "much of whatever success the Team has enjoyed to this date is because of its and DSI's recognition of the need for maintenance of a continuing day-to-day intimate working relationship between American and Turkish technicians and officials."[105] The team eventually used the Yesilirmak River basin as a kind of test case for the transfer of knowledge and skills in river basin planning. Throughout this period of direct Bureau training and engagement with DSI, which lasted well into the late 1960s, the State Department continuously expressed support for Turkey as a reliable geopolitical ally in a critical region. The United States also used its influence to assist the Turkish government to uncover financing for a series of large-dam projects, including substantial hydroelectric projects in the Seyhan and Yesilirmak River basins. Turkey's interest in hydropower development as a key element of its economic and geopolitical security aims continues to the present day. The numerous hydroelectric schemes and river basin development initiatives implemented in the decades since the Bureau advisory team ended its engagement have fomented socioecological disruptions and prompted a significant anti-dam social movement.[106]

South America, Central America, and the Caribbean

Argentina

The Bureau never played a significant role in water resource development in Argentina, but hydroelectric development there nevertheless became a concern of the State Department in the 1960s due to the Argentine government's development assistance agreements with the Soviet Union and the perception that American companies were being treated unfairly in bids for construction work on large-dam projects. Argentina has a long history of hydroelectric development—its first hydroelectric power station was built in 1898 on the Rio Primero[107]—and the Bureau's Denver office followed water resource development trends in the country with interest in the 1930s. The State Department traced debates over financing of the El Chocón hydroelectric dam on the Limay River in the early 1970s, mostly due to concerns expressed by US embassy officials that a British firm was planning to develop a joint proposal with a Russian venture to finance part of the project.[108] First proposed in 1966, the El Chocón Dam was seen by the Argentine state as a crucial component of developing Patagonia, and the dam was eventually completed in 1973. The other major hydroelectric project that piqued the interest of US officials was the Salto Grande Dam, an example that clearly delineates the close linkage between US geopolitical and economic interests in the sphere of large-dam production. In early 1973 the vice president of Boston-based Charles T. Main International, a prominent American engineering firm, approached the US embassy staff in Buenos Aires, expressing concern over "irregularities" in the bidding process for a contract as consulting engineer on the hydroelectric plant associated with the massive Salto Grande Dam, located on the Uruguay River on the Argentina-Uruguay border. Main did not ask for US "intervention," but wanted to determine whether the bidding process might be reconsidered.[109] Embassy officials later noted that the awarding of the contract to a Canadian firm, Acres International, was unfortunate due to the "dangerous and disadvantageous" technical features of the project design and was "detrimental to US interests."[110] After numerous visits and cajoling by both Main representatives and US officials, the Argentine and Uruguayan governments reversed their decision and awarded the contract to Main, bringing to a murky conclusion what one official called "this important and bitterly fought-over phase of the project."[111] Construction of the Salto Grande project, at the time one of the world's largest hydroelectric projects with a generating capacity of nearly 1,900 megawatts, was finished in 1979.

Bolivia

In 1962 the Inter-American Development Bank requested assistance from the Bureau of Reclamation in "clearing up certain technical difficulties" centered on the Villa Montes project in Tarija Province of Bolivia, a large-scale irrigation scheme expected to increase food production in the region. The Bureau dispatched one engineer, Arthur Johnson, to examine the issue, and he concluded that additional data on various hydrological processes would be necessary to address the difficulties.[112]

Brazil

A headline in the June 19, 1943, issue of *Business Week* read, "Brazil Plans Its Own TVA." The article that followed proceeded to lay out the Brazilian government's plans to use the waters of the São Francisco River basin as a basis for industrialization and expanded agricultural production. The 260,000-square-mile basin, beginning in the mountainous region just north of Rio de Janeiro and draining north and east to the Atlantic Ocean, was identified by a US economic mission to Brazil in the early 1940s as the key to regional development, given the presence of significant hydropower and irrigation potential, a large population, and important mineral reserves. US engineering firms were key actors in the development of two of the basin's early hydroelectric projects. Materials for the World Bank–financed Paulo Affonso Falls Dam—completed in 1955—on the main channel were provided by Westinghouse International Electric Company, while a consortium of US firms, including the San Francisco–based International Engineering Company, assisted in design and construction of the Três Marias Dam—completed in 1962.[113]

These initial developments set the stage for two of the Bureau's comprehensive river basin assessments in Brazil in the 1960s: a report on the Araguaia-Tocantins River basin in the central part of the country, and a detailed survey of the São Francisco basin. Both studies were commissioned by USAID and were connected to the Kennedy administration's Alliance for Progress initiative in the Americas. Somewhat remarkably, the three-person team examining the Araguaia-Tocantins basin based its recommendations for the 750,000-square-kilometer region on three months of reviewing relevant data and "field reconnaissance by airplane, boat and car."[114] Not surprisingly, it concluded that a "properly planned resource development program" could propel the "present extractive, subsistence-level, frontier-type local economy" of the basin ("among the major potential areas of development in Brazil and in the world") into "a major, self-liquidating contributor" to national development.[115] During this same period the Bureau also became involved in similar reconnaissance studies of other river basins in the country, including the Rio Grande and Piranhas river basins in northeastern Brazil, in addition to inspections of specific dam projects.

The Bureau's work in the São Francisco basin was more intensive, involving a team of nine people whose stays extended for as long as two years in some cases. The mission was broader as well, encompassing, in addition to the usual documenting of land and water resource development potential, a specific training component for Brazilian "professional and administrative personnel" in water resources investigations and suggestions regarding institutional changes in the government that would facilitate more effective river basin development.[116] The report's conclusions regarding the basin's development potential were ambitious: it detailed over a dozen major multi- and single-purpose irrigation, hydroelectric, and flood control projects for the basin's main channel and tributaries, and it argued that development of the São Francisco basin was a national imperative. Many of the proposed projects have since come to fruition, and at present the São Francisco is one of the most heavily controlled river systems in South America, with attendant social and environmental disruptions accompanying basin alteration.

At a geopolitical level, US interest in providing technical assistance to Brazil was in part prompted by concerns that the Brazilian government might turn toward Moscow for development funding. A somewhat obscure and highly classified 1962 message from the Central Intelligence Agency (CIA) in 1962 revealed a conversation in which Brazilian president João Goulart suggested to Soviet premier Nikita Khrushchev that the "soviets announce and begin several spectacular investment projects . . . such as building a metro or hydroelectric plant."[117] This conversation may have simply been an effort on the part of Goulart to leverage additional assistance from the American government.

Chile

Bureau activities in Chile were minimal, although a pair of Bureau experts (a seismologist and an irrigation specialist) journeyed to the Elqui Valley region of Chile as consultants over the period 1952–1954. In comparison with many of the Bureau's other overseas endeavors, the consultants' work appears to have been purely technical and largely training-oriented. It involved, for example, site visits to projects where the expert delivered a seminar on "groundwater geology and hydrology and drilling methods and practices" to Chilean engineers.[118]

Colombia

In Colombia, the Bureau's activities were limited to a brief assignment carried out by two specialists who produced a report on the land and water resources of the Magdalena-Cauca River basin, which encompasses 70 percent of the population of the country and is the site of several large water development projects. The goal of the Bureau operation—carried out under the auspices of USAID—was "to conduct a reconnaissance-type examination" of the basin.[119] The report, published in 1967, concluded that the basin "constitutes the single most important segment of Colombia in terms of present development and expansion potentials for immediate utilization," and that its resource development potential was significant in terms of hydroelectric and irrigation development.[120] Several years later, in 1972, an American engineering firm requested that US embassy officials in Bogota investigate a contract for the construction of a hydroelectric station associated with the Chivor Dam. In a situation parallel to that of Charles T. Main in Argentina, the firm, BECM International Corporation, was not awarded the contract despite being the low bidder, an occurrence US officials attributed to BECM's reputation as a "not respectable" operation.[121] These and other similar communications concerning the interests of American firms in the construction of hydroelectric projects throughout the world demonstrate the tight links between US foreign policy and American business interests.

Costa Rica

The Bureau's primary engagement in Costa Rica was a modest investigation of the development potential of the Tempisque Valley, which drains into the Pacific Ocean in the northwestern part of the country, carried out under the auspices of the Point Four program. Bureau specialists carried out this investigation over two brief visits in 1951 and 1952, and its central goal involved an assessment of the basin's irrigation potential.[122] The Costa Rican government also pursued an ambitious hydroelectric development strategy during the 1950s and 1960s, including the La Garita Dam on the Rio Grande de Tarcoles near the capital of San Jose.[123] In the planning phases of these projects, US embassy officials often played the role of interlocutor between Costa Rican officials and potential funding agencies such as the World Bank. Still, the Costa Rican government's overtures to the Soviet Union in the early 1970s to finance another hydroelectric project, the Boruca Dam on the Terraba River, generated concerns among State Department officials that Costa Rican–Soviet relations were improving and that the result would be a "significant increase in the Soviet presence here."[124] Geopolitics and the construction of large dams continued to be intertwined.

Dominican Republic

The Bureau's official engagement with river basin planning in the Dominican Republic was fairly minimal, consisting of a 1974 investigation of the water resource development potential of the Yaque del Norte River basin brokered through the Inter-American Development Bank (IDB). A single Bureau engineer carried out a three-day "field appraisal" and, in addition to confirming the feasibility of an ongoing irrigation project, outlined a series of data collection and planning activities that needed to be accomplished in advance of any substantive undertakings.[125] The keystone project for development of the Yaque del Norte basin was the Tavera Dam, a major hydroelectric scheme that had been conceived by planners in the Dominican government in the late 1950s and was erected over the period 1969–1973. The evolution of the Tavera project was documented by US embassy officials in Santo Domingo who negotiated financial support for the dam from the IDB because of the US government's close ties to the regime of Joaquin Balaguer. The project, the largest dam in the Dominican Republic, is also notable because of the "national movement" that emerged in support of the dam. As funding for the project appeared to be unlikely in the late 1960s, a coalition of diverse social groups—including "campesinos, industrialists, professionals, and Dominicans residing in New York," as well as the leadership of the armed forces, the police, and the Catholic Church—all expressed unequivocal support for the project and even organized grassroots financial contributions from their respective collectives. American officials marveled at the emblematic power of large dams, noting that the "fact that a national movement of this nature could arise in a country so divided internally" was a tribute to the Tavera Dam as a symbol "of economic progress and development."[126] Operation of the Tavera Dam and two other dams on the Yaque del Norte River (site of the Bureau's initial water resources survey) has been greatly diminished by enormous sedimentation rates—the Tavera project functions at only 40 percent efficiency today, and its reservoir will soon be half filled with sediment—associated with deforestation and deleterious land use practices in the basin.[127]

Haiti

Haiti's single large multipurpose dam, the Peligre, located on the Artibonite River, was completed in 1956 as a water storage project for irrigation, and a hydroelectric plant was added in 1971. There is remarkably little information on the origins of the dam and its eventual construction, and the details of US involvement in the planning of the dam and the broader vision of river valley development are difficult to trace. However, it is clear that the US government played a substantive role in the dam's creation and considered it an element of larger geopolitical aims concerning Haiti. In the early 1950s the Haitian government created the Artibonite Valley Development Organization—modeled after the TVA—to coordinate development planning for the basin, which had a long history of irrigation dating back to the French colonial period. A report by a Haitian civil engineer in 1956 noted that although valley development was to be primarily an irrigation project, the "necessity of including hydroelectric power in this project is dictated not only by engineering considerations but also by economic and social conditions."[128] All investigations prior to construction of the dam were carried out by the American construction firm Knappen Engineering Company and were facilitated by a technical cooperation agreement authorized by the United States (under the auspices of the Institute of Inter-American Affairs and the Foreign Operations Administration) and Haitian governments in 1951. Eventually Haiti received a US$14 million loan from the US Export-Import Bank to go ahead with the project, and the dam was erected by the US firm Brown & Root.

In subsequent years, the Haitian state's plan to finalize the Peligre Dam's hydroelectric facility in the 1960s became complicated after the United States cut off economic and military assistance in the early 1960s because of the "corruption and repressive regime" of François "Papa Doc" Duvalier.[129] Despite this public stance, State Department officials continued to contemplate less visible means of supporting the Duvalier regime due to overarching geopolitical considerations. As laid out by embassy officials in 1965, the "basic US policy goal . . . remains entirely valid, i.e. maintenance effective USG presence . . . in Haiti so as to be in position influence events when Duvalier passes from scene, and to frustrate any Communist attempt to take over before or after."[130] Despite consistent pleas from the Duvalier regime to the United States for financial assistance in building the power plant, the hydroelectric facility was only completed in 1971 with minimal US assistance. Ultimately, the dam originally constructed with Export-Import Bank funding became a social and environmental disaster, beset by dam-induced deforestation, impoverishment and subsequent out-migration of valley communities, and sediment accumulation so severe it has effectively reduced the dam's lifetime as a power producer by 50 percent.[131] Plans to rehabilitate the decrepit hydroelectric facility—operating at half its capacity—were bolstered by the recent announcement of a US$20 million grant to Haiti from the Inter-American Development Bank (IDB) that is linked to an additional US$40 million from other sources committed to the restoration project.[132]

Honduras

The Bureau sent a single engineer to Honduras for one month in 1951 to ascertain the feasibility of hydroelectric development in several river basins throughout the countryside, although the Honduran government was most interested in continuing work on a hydroelectric project on the Rio Lindo, a tributary of the Ulua River in the northwestern region. In a clear example of the political-economic underpinnings of technical assistance, the United Fruit Company, whose representatives had wanted for several years to develop the water resources of the Rio Lindo for both hydropower and irrigation benefit, provided economic and engineering expertise to the visiting Bureau staff member.[133] Several American firms (including Morrison Knudsen) became involved in the construction of the Rio Lindo hydroelectric facility, built in several stages from the early 1960s to 1971.

Mexico

While direct Bureau involvement in Mexico during the Cold War era covered in these pages was negligible, there is no question that reclamation activities in the United States influenced patterns of state-directed water resource development in Mexico, particularly in the country's arid regions. The most straightforward linkage between Bureau experts and the Mexican government occurred in the 1920s. During that decade a number of the "Bureau's best personnel" procured leaves of absence in order to provide technical consultations to Mexican officials involved in their nation's nascent water development program. A significant outcome of these collaborations was the inception in 1926 of the National Irrigation Commission, an agency that, like water bureaucracies in Turkey, Ethiopia, and many other states, patterned its organizational structure after the Bureau and sent dozens of its staff to Denver for technical and administrative training.[134]

Nicaragua

The Bureau's relatively modest engagements in Nicaragua spanned several decades but are most notable for the audacity of the river basin alterations its engineers envisioned in the 1970s, almost none of which came to fruition. As part of the newly minted Point Four program initiatives, a Bureau mission headed by Robert Newell spent five weeks conducting a reconnaissance survey of all potential multipurpose water resource development locations in the country in the spring of 1951. Newell observed that although the "economy of the country is based principally on agriculture" and the "water resources are almost completely undeveloped," development of hydroelectric power appeared to be the chief interest or at least "the first interest of Nicaraguan officials." Clearly the electricity-generating function of large dams had captured the imagination of the Nicaraguan state at the time.[135] The final report—like several Bureau documents concerning its overseas activities never made publicly available—outlined a series of potential hydroelectric dams that could be constructed on the Tipitapa River near Managua and within the Tuma/Grande River basin that empties into the Atlantic Ocean. The report also mentioned the feasibility of a series of rather complex inter-basin diversion projects that would connect Lake Nicaragua to the Pacific Ocean and greatly expand the power and irrigation possibilities of the Lake Managua drainage area.[136] While geopolitical concerns over Nicaragua had not yet become the focal point of US relations with Central America that they would become during the Sandinista revolutionary period in the late 1970s,[137] official US policy focused, as it had since the 1930s, on support of the Somoza dynasty, and the Bureau received communiqués from the US ambassador in Managua urging the "need of speed" in pressing ahead with hydrological investigations.[138] Newell's report was followed a year later by more thorough hydrological investigations of potential dam sites on the Tuma River, which had by then become a priority for Nicaraguan officials. The Mancotal (also called Rio Tuma) Dam, a rather modest 25-megawatt hydroelectric scheme constructed on the Tuma between 1961 and 1966, was funded with a US$12 million IBRD loan and an additional US$2.5 million from the US government.

In the mid-1970s a three-person Bureau team conducted research in Nicaragua to evaluate the feasibility of an ambitious scheme to reconfigure nearly all the nation's river systems in a way that, if realized, would have transformed Nicaragua's waterways into a vast, interconnected series of canals and reservoirs in the service of irrigation and hydroelectric development. The plan, known as *Hacia La Meta* ("toward the goal"), was conceived by a "group of Nicaraguan engineers" under the guidance of the IDB, and the IDB suggested Bureau involvement in early 1976. As noted in the Bureau report, the concept "envisions a series of dams on the several large rivers flowing to the Atlantic Ocean" that would make it possible, through "means of connecting irrigation canals, pipes, and rivers" to divert this water to the more populous Pacific region to meet its "agriculture, domestic, municipal, and industrial needs." The westward "journey" of the water would also facilitate the generation of "hydroelectric power at several elevation drops," and the immense reservoir surface areas would "provide a navigable waterway to the interior of the country" and ultimately connect Lake Nicaragua to the Atlantic coast.[139] A key motivation of the plan was supposedly to "unite the nation into a single social entity, rather than three distinct social regions," wherein the isolated eastern regions would be "opened to development" under the auspices of water resource exploitation.[140] Although the Bureau team expressed great enthusiasm for the grandiose scheme—the report concludes that the "economic future of Nicaragua probably depends on the Hacia La Meta Concept"[141]—none of its major projects have, to my knowledge, been completed, and Nicaragua's internal political turmoil during the decades that followed, exacerbated by US foreign policy decisions, unquestionably placed the plan on a back burner. However, construction of the long-delayed 250-megawatt Tumarin Dam on the Rio Grande, financed through Brazil's well-known Eletrobras utility company at a total cost of over US$1 billion, and seemingly a descendant of one of the Hacia La Meta projects, is scheduled to begin later in 2014.[142]

Peru

In 1961 Bureau engineer James Knights traveled to Peru for three months at the request of the government to consult on improvements and enhancement to two existing irrigation projects in the central part of the country and to offer a preliminary evaluation of the potential for hydroelectric development.[143] Several years later the State Department, through its emissaries in the US embassy in Lima, became deeply embroiled in largely hidden negotiations over financing of an immense hydropower complex, with a production potential of 1,100 megawatts, on the Mantaro River, which flows out of the eastern slopes of the Andes. The Americans' central interest revolved around whether the Peruvians would request financing through the World Bank and USAID in lieu of other sources for the nearly US$500 million scheme. After a period of "public debate and acrimonious arguments" in 1966, the Peruvian government canceled an agreement with an Anglo–West German consortium originally tendered to construct the project, and at some point the government asked embassy officials if they could recommend a "reputable" consulting firm to review the project contracts. Again illustrating how technopolitics at multiple scales infiltrates large dams and their networks, American officials concluded that the "Mantaro imbroglio" was the result of "internal political considerations" related to the project's strong identification with a single Peruvian political party. Moreover, many Peruvians and the World Bank considered the project "unnecessary and ill-advised" given the country's scarce financial resources.[144] Domestic conditions shifted in later years, however, and at present the Mantaro River is almost completely regulated via a series of hydroelectric and irrigation dams.

Uruguay

In another of the numerous studies instigated in South and Central America in the early 1950s under the umbrella of the Point Four program, the Bureau conducted a very specific evaluation of a hydroelectric power site at Rincon de Baygorria on the Rio Negro in Uruguay over one month in early 1954.[145] In the following decade, US embassy officials monitored Uruguayan participation in development of the Salto Grande Dam, on the Uruguay River shared with Argentina (see under "Argentina" above), primarily due to the comments of a visiting Russian professor who indicated that the "Soviet Union would have no difficulty granting financial and technical assistance for the Salto Grande if Argentina and Uruguay so requested."[146]

Venezuela

Venezuela's construction of the Guri Dam, a large hydroelectric scheme on the Caroni River in the eastern region of the country initiated in 1963, is a further example of the tight connections between the State Department and American commercial interests. As the project progressed in the mid-1960s, US embassy officials stayed in close contact with Edgar Kaiser and other representatives of Kaiser Engineers and Constructors regarding problems associated with the project's implementation. Facing construction work that was thirteen months behind schedule and huge cost overruns, Kaiser hoped the US government would facilitate positive requests to both the Venezuelan government and the World Bank to provide additional funding, given that "working capital for this stage of the operation has risen from an estimated $6 million to over $30 million." A number of problems confounded the dam's construction, including labor problems and an inaccurate survey of the site's geological conditions that led to additional earth moving.[147] The financial woes were eventually ironed over, and the dam was inaugurated in 1968, in a ceremony that included an impressive array of representatives from the world of diplomacy, international finance, and the engineering profession, many if not most of whom were American citizens.[148]

NOTES

CHAPTER ONE

1. Eric Hobsbawm, *The Age of Extremes: A History of the World, 1914–1991* (New York: Vintage Books, 1994), 13.
2. While understandings of what constitutes a "large" dam have changed over time, I will employ the definition offered by the International Commission on Large Dams (ICOLD)—a dam of over 15 meters in height or one storing over 3 million cubic meters of water in its reservoir—throughout this work. See ICOLD, *Dams and the World's Water* (Paris: ICOLD, 2007), 28.
3. World Commission on Dams (WCD), *Dams and Development: A New Framework for Decision-Making: A Report of the World Commission on Dams* (London: Earthscan, 2000).
4. Ibid., 310.
5. C. D. Thatte, "Aftermath, Overview and an Appraisal of Past Events Leading to Some of the Imbalances in the Report of the World Commission on Dams," *Water Resources Development* 17 (2001): 343–51; John Briscoe, "Overreach and Response: The Politics of the WCD and Its Aftermath," *Water Alternatives* 3 (2010): 399–415.
6. This may explain in part why more than ten years after its release, the WCD report has had a seemingly minor impact on the continuing debates over large dams and their socioecological consequences. See Deborah Moore, John Dore, and Dipak Gyawali, "The World Commission on Dams +10: Revisiting the Large Dams Controversy," *Water Alternatives* 3 (2010): 3–13.
7. See Wiebe Bijker, "Dikes and Dams, Thick with Politics," *Isis* 98 (2007): 109.
8. This is a conservative estimate, and a majority of these displaced persons were forcibly removed. See Patrick McCully, *Silenced Rivers: The Ecology and Politics of Large Dams*, 2nd ed. (London: Zed Books, 2001), 7–8.
9. Brian D. Richter et al., "Lost in Development's Shadow: The Downstream Human Consequences of Dams," *Water Alternatives* 3 (2010): 14–42.
10. See Charles J. Vörösmarty et al., "The Storage and Aging of Continental Runoff in Large Reservoir Systems of the World," *Ambio* 26 (1997): 210–19; Charles J. Vörösmarty et al., "Anthropogenic Sediment Retention: Major Global Impact from Registered River Impoundments," *Global and Planetary Change* 39 (2003): 169–90; Francis J. Magilligan and Keith H. Nislow, "Changes in Hydrologic Regime by Dams," *Geomorphology* 71 (2005): 61–78; and N. Leroy Poff et al., "Homogenization of Regional River Dynamics by Dams and Global Biodiversity Implications," *Proceedings of the National*

Academy of Sciences of the United States of America 104 (2007): 5732–37. I intentionally cite the biophysical research in this note to underscore the scientific consensus around the profound material effects of large dams. One geophysicist calculated that the amount of water stored in reservoirs has had a measurable effect on the earth's rotation and gravitational field. See Benjamin Fong Chao, "Anthropogenic Impact on Global Geodynamics Due to Reservoir Water Impoundment," *Geophysical Research Letters* 22 (1995): 3529–32.

11. It is important to note that not all large dams include concrete as the primary construction material; nor do the different types of concrete dams—a category that includes gravity, buttress, arch, and gravity-arch forms—use a universal mixture of concrete. Each dam is unique to the particular characteristics—river flows, geology, and climate—of its site.

12. Richard Lee Strout, "Kilowatts for the Lamps of China," *New Republic*, March 26, 1945, 411.

13. See, for example, John H. Perkins, *Geopolitics and the Green Revolution: Wheat, Genes, and the Cold War* (New York: Oxford University Press, 1997); and Nick Cullather, *The Hungry World: America's Cold War Battle against Poverty in Asia* (Cambridge, MA: Harvard University Press, 2010).

14. For a general treatise on modernization schemes and their ideological roots, see James Scott, *Seeing like a State: How Certain Schemes to Improve the Human Condition Have Failed* (New Haven, CT: Yale University Press, 1998). For analysis of the role of technology in promoting a specifically American form of modernization, see Michael Adas, *Dominance by Design: Technological Imperatives and America's Civilizing Mission* (Cambridge, MA: Belknap Press of Harvard University Press, 2006). The connections between Cold War geopolitics and large dams were perhaps most famously played out over funding and eventual construction of the Aswan High Dam on the Nile River in Egypt during the late 1950s and early 1960s. For an illuminating and highly detailed account of the geopolitical and economic rationalizations of the Eisenhower administration for "killing" any American assistance to Egypt for construction of the dam (and its eventual construction via Soviet financial aid), see Guy Laron, *Origins of the Suez Crisis: Postwar Development Diplomacy and the Struggle over Third World Industrialization, 1945–1956* (Baltimore: Johns Hopkins University Press, 2013), 136–53.

15. See especially David Ekbladh, *The Great American Mission: Modernization and the Construction of an American World Order* (Princeton, NJ: Princeton University Press, 2010).

16. A spate of recent work by historians and historically minded social scientists has partially traced the evolution of the river basin as the key modality of water resource planning and development. See: François Molle, "River-Basin Planning and Management: The Social Life of a Concept," *Geoforum* 40 (2009); Daniel Klingensmith, *"One Valley and a Thousand": Dams, Nationalism, and Development* (New Delhi: Oxford University Press, 2007); and James Wescoat, "'Watersheds' in Regional Planning," in *The American Planning Tradition: Culture and Policy*, ed. Robert Fishman (Washington, DC: Wilson Center, Smithsonian Institution, 2000).

17. See Bureau of Reclamation, "Summary of Foreign Activities Program," Staff Information Letter, November 2, 1959, 1, Record Group 115 (hereafter RG 115), International Affairs Reports, Foreign Governments (1951–1978), box 2, National Archives and Research Administration—Rocky Mountain Region (Denver, CO) [hereafter NARA (Denver)].

18. I use "tricontinental" throughout the book to denote the regions of the planet more typically clustered under the terms "Third World," "Global South," "underdeveloped,"

and so on. I follow the rationale of Robert Young, who sees "tricontinental"—while unwieldy—as perhaps the least problematic category for the lands and peoples of Asia, Africa, and Latin America. This preference resides in part due to its adoption at the 1966 inaugural conference of the Organization of Solidarity of the Peoples of Africa, Asia, and Latin America and subsequent anti-imperialist sensibilities in the journal *Tricontinental*. See Robert J. C. Young, *Postcolonialism: An Historical Introduction* (Oxford: Blackwell, 2001), 4–5.

19. The most measured study of the positive and negative impacts of large dams is most likely WCD's *Dams and Development*. Currently, the most vociferous support for the continuing relevance of large dams as a source of relatively clean energy and a vital means of storing water for a variety of human uses comes from the dam industry itself (including, prominently, several Chinese dam-building agencies and corporations) and select international financial institutions. For example, see ICOLD, *Dams and the World's Water*, and World Bank Group, *Directions in Hydropower* (Washington, DC: World Bank Group, March 2009). Critical studies of large dams and their multitude of social and ecological impacts include Edward Goldsmith and Nicholas Hildyard, *Social and Environmental Effects of Large Dams*, vol. 1, *Overview* (San Francisco: Sierra Club Books, 1984); Edward Goldsmith and Nicholas Hildyard, eds., *Social and Environmental Effects of Large Dams*, vol. 2, *Case Studies* (Cornwall, UK: Wadebridge Ecological Centre, 1986); William Adams, *Wasting the Rain: Rivers, People and Planning in Africa* (Minneapolis: University of Minnesota Press, 1992); Fred Pearce, *The Dammed: Rivers, Dams, and the Coming World Water Crisis* (London: Bodley Head, 1992); William L. Graf, "Dam Nation: A Geographic Census of American Dams and Their Large-Scale Hydrological Impacts," *Water Resources Research* 35 (1999); McCully, *Silenced Rivers*; and Sanjeev Khagram, *Dams and Development: Transnational Struggles for Water and Power* (Ithaca, NY: Cornell University Press, 2004).

20. I employ "genealogy" as a "historical narrative that explains an aspect of human life by showing how it came into being." See Mark Bevir, "What Is Genealogy?," *Journal of the Philosophy of History* 2 (2008): 263. In contrast to Bevir's interpretation, I emphasize the nonhuman (e.g., technological, biophysical) in addition to the human, as well as the connections between the two. Genealogies (at least in the sense considered by Michel Foucault and scholars building on his work) should not be considered as historical moments that are sequenced along a temporal pathway, but rather as analytical reference points that assist in delineating the political rationalities that define certain phenomena. See Wendy Larner and William Walters, "The Political Rationality of 'New Regionalism': Toward a Genealogy of the Region," *Theory and Society* 31 (2002): 394–95.

21. It would be impossible to explain the dissemination of large dams without reference to the parallel rise of technologies of electricity production and transmission and the economic and political relations prompted by this novel form of energy. For a seminal examination of the early evolution (late nineteenth and early twentieth century) of electricity systems, see Thomas Hughes, *Networks of Power: Electrification in Western Society, 1880–1930* (Baltimore: Johns Hopkins University Press, 1983).

22. I am tempted to refer to this work and its subject matter as a meditation on "things without history," which is certainly a tribute to Eric Wolf's monumental work. But I am reticent because of the sheer overconfidence that explicitly rears its head in such a formulation, and also because it does a disservice to the millions of people to whom Wolf's title refers, largely ignored by Eurocentric histories. Still, most of the large dams of the twentieth century are indeed "things without history," and one of

my goals is to recover this history and to think through why it matters. See Eric Wolf, *Europe and the People without History* (Berkeley: University of California Press, 1982).

23. For an extended discussion from one of actor-network theory's primary—but not, as is commonly assumed, only—architect, see Bruno Latour, *Reassembling the Social: An Introduction to Actor-Network-Theory* (Oxford: Oxford University Press, 2005). I make use of the term "network" throughout this chapter, but Latour's overview of this seminal concept in science and technology studies, which encompasses the rich and contested history of the network idea and explores the numerous interdisciplinary efforts to come to grips with actor-networks as ontological, epistemological, and methodological entities, is beyond my desired scope. For a perspective on actor-network theory's adoption within human geography, see Noel Castree, "False Antitheses? Marxism, Nature and Actor-Networks," *Antipode* 34 (2002); and for my own interpretation of the mutual construction of social and biophysical networks, mindful of the need to account for power effects and spatial scale within these constructions, see Chris Sneddon, "Reconfiguring Scale and Power: the Khong-Chi-Mun Project in Northeast Thailand," *Environment and Planning A* 35 (2003).

24. Manuel DeLanda, building on and extending the ideas of philosopher Gilles Deleuze, portrays assemblage thinking as an ontological position emphasizing the hybridity of human, technological, and ecological entities extending across space and time. A key stance of DeLanda, and one that I share, is that assemblages are "wholes whose properties emerge from the interactions between parts," and that investigating only the parts of the assemblage will omit important questions. See Manuel DeLanda, *A New Philosophy of Society: Assemblage Theory and Social Complexity* (London: Continuum, 2006), 5. Additionally, the assemblage metaphor can be seen as an "orientation" toward research that highlights the "ontological diversity of agency"—not falling into the trap of seeing it exclusively in human terms. Large-dam assemblages thus become participants, however unwitting, in the geopolitical and technological processes that created them. See Colin McFarlane and Ben Anderson, "Thinking with Assemblage," *Area* 43 (2011): 162.

25. Stephen J. Collier and Aihwa Ong, "Global Assemblages: Anthropological Problems," in *Global Assemblages: Technology, Politics, and Ethics as Anthropological Problems*, ed. Aihwa Ong and Stephen J. Collier (Malden, MA: Blackwell, 2005), 4–5.

26. Ben Anderson and Colin McFarlane, "Assemblage and Geography," *Area* 43 (2011): 124.

27. I borrow the phrase "technopolitical network" from the insightful work of Samer Alatout, who traces the evolution of networks of Zionist political agents, technical expertise, environments, and discourses revolving around debates over water abundance in the decades before the formation of the Israeli state in 1948. See Samer Alatout, "Bringing Abundance into Environmental Politics: Constructing a Zionist Network of Water Abundance, Immigration, and Colonization," *Social Studies of Science* 39 (2009): 363–94. Much of the challenge in distinguishing between "assemblage" and "network" is articulating the rather different origins of the two terms within theoretical traditions that are now widely used in geography and related disciplines. Jonathan Murdoch, a key figure in advocating associational thinking within the social sciences, at times uses the two phrases interchangeably; see J. Murdoch, "Ecologising Sociology: Actor-Network Theory, Co-construction and the Problem of Human Exemptionalism," *Sociology* 35 (2001). A key point is that both emphasize the problematic separation of human and nonhuman (the latter includes both technologies and ecological things and processes) and try to highlight not just the relationships, but the absolute inseparability—materially and analytically—of the human and nonhuman spheres. My aim in this book is to nod to the important theoretical

discussions around the terms without overwhelming the nonspecialist reader with unproductive jargon.

28. Murdoch, "Ecologising Sociology," 119–20.

29. Bruce Braun, "Environmental Issues: Global Natures in the Space of Assemblage," *Progress in Human Geography* 30 (2006): 644. Recent initiatives in human-environment geography emphasize the need for recognition of the "global assemblages" of political and economic institutions, development interventions, biophysical systems, and cultural interpretations of "nature" that influence and direct ecological transformations. See Laura Ogden et al., "Global Assemblages, Resilience, and Earth Stewardship in the Anthropocene," *Frontiers in Ecology and the Environment* 11 (2013): 341–47. Previous work in science and technology studies and associated fields has explored the relations between water and politics as mediated by specific technologies and forms of knowledge in a way that resonates with the understanding of large dams as assemblages put forward here. See, for example, Marianne de Laet and Annemarie Mol, "The Zimbabwe Bush Pump: Mechanics of a Fluid Technology," *Social Studies of Science* 30 (2002); Rupali Phadke, "Assessing Water Scarcity and Watershed Development in Maharashtra, India: A Case Study of the Baliraja Memorial Dam," *Science, Technology, & Human Values* 27 (2002); Tsegaye H. Nega, "Saving Wild Rice: The Rise and Fall of Nett Lake Dam," *Environment and History* 14 (2009); Alatout, "Bringing Abundance into Environmental Politics"; and Jessica Barnes, "Managing the Waters of Ba'th Country: The Politics of Water Scarcity in Syria," *Geopolitics* 14 (2009): 510–30.

30. Indeed, as several observers interested in the history of things and in how human and nonhuman entities become entangled and sutured together as "quasi-objects" have emphasized, there is a pressing need to historicize the production of assemblages or associations. For example, geographers stress the seeming lack of interest within science studies in explaining how and why networks of humans and nonhumans are constructed and remain more or less stable over time. See Karen Bakker and Gavin Bridge, "Material Worlds? Resource Geographies and the 'Matter of Nature,'" *Progress in Human Geography* 30 (2006): 17.

31. Of course, power is relational and "makes its presence felt through a variety of modes playing across one another" as it is exercised over time and across space. See John Allen, *Lost Geographies of Power* (Malden, MA: Blackwell, 2003), 196.

32. Jonathan Murdoch and Terry Marsden, "The Spatialization of Politics: Local and National Actor-Spaces in Environmental Conflict," *Transactions of the Institute of British Geographers* 20 (1995): 372.

33. See DeLanda, *New Philosophy*, 40–46; and John Allen, "Powerful Assemblages?," *Area* 43 (2011): 155.

34. In an effort to demonstrate the continuing relevance of Marxist approaches to understanding human-nature-technology relations, Kirsch and Mitchell argue compellingly that "there is social intentionality in turning relationships into things; there are reasons for putting networks together, even if those reasons themselves are highly structured by and determined within the contested relationships that constitute capitalism as a social totality." See Scott Kirsch and Don Mitchell, "The Nature of Things: Dead Labor, Nonhuman Actors, and the Persistence of Marxism," *Antipode* 36 (2004): 700. While I would qualify this stance by accentuating the importance of geopolitical calculation that partially transcends the imperatives of capitalist economic development, I find it hard to disagree with this general perspective.

35. While I resist cheerleading for any particular discipline, the field of geography, with its tradition of a critical and interdisciplinary approach to knowledge production, its insis-

tence on the importance of place, and its emphasis on the social construction of both spatial scale and human-environment relations, is an efficacious vehicle for this book.

36. What I offer here are neither precise definitions nor sweeping reviews of established disciplinary territories. Rather, my aim is to lay out those conceptual regions—in this case historical research—that directly complement the book's broader aims. To that end, there is a great deal of work that I must regrettably exclude from consideration.

37. For a recent statement on the purview of environmental history, see J. Donald Hughes, *What Is Environmental History?* (Cambridge: Polity Press, 2006).

38. See, for salient examples, Donald Worster, *Rivers of Empire: Water, Aridity, and the Growth of the American West* (New York: Pantheon Books, 1985); Ted Steinberg, *Nature Incorporated: Industrialization and the Waters of New England* (Amherst: University of Massachusetts Press, 1991) and Richard White, *The Organic Machine* (New York: Hill and Wang, 1995).

39. For a comprehensive overview, see Jeffrey Stine and Joel Tarr, "At the Intersection of Histories: Technology and the Environment," *Technology and Culture* 39 (1998): 601–40.

40. Stine and Tarr, "Intersection of Histories," 625–26.

41. One prominent environmental historian goes so far as to argue that his US brethren in the field "are exceptional . . . in their reluctance to confront American engagement with the rest of the world." See J. R. McNeill, "Observations on the Nature and Culture of Environmental History," *History and Theory* 42 (2003): 17. There are signs, however, that this position is changing. A recent pathbreaker in this regard is J. R. McNeill and Corinna Unger, eds., *Environmental Histories of the Cold War* (Cambridge: Cambridge University Press, 2010).

42. Several recent studies are working in a similar vein. See David Biggs, "Reclamation Nations: The U.S. Bureau of Reclamation's Role in Water Management and Nation-Building in the Mekong Valley, 1945–1975," *Comparative Technology Transfer and Society* 4 (2006); Richard P. Tucker, "Containing Communism by Impounding Rivers: American Strategic Interests and the Global Spread of High Dams in the Early Cold War," in *Environmental Histories of the Cold War*. Ed. McNeill and Unger; Ekbladh, *Great American Mission*; and Heather Hoag, *Developing the Rivers of East and West Africa: An Environmental History* (London: Bloomsbury, 2013).

43. For illuminating examples of the "new historicism" put into practice, see Michael E. Latham, *Modernization as Ideology: American Social Science and "Nation Building" in the Kennedy Era* (Chapel Hill: University of North Carolina Press, 2000); Robert J. McMahon, "Introduction: The Challenge of the Third World," in *Empire and Revolution: The United States and the Third World since 1945*, ed. Peter L. Hahn and Mary Ann Heiss (Columbus: Ohio State University Press, 2001), 1–14; Nils Gilman, *Mandarins of the Future: Modernization Theory in Cold War America* (Baltimore: Johns Hopkins University Press, 2003); David Engerman, "The Romance of Economic Development and New Histories of the Cold War," *Diplomatic History* 28 (2004): 23–54; and, most comprehensively, Odd Arne Westad, *The Global Cold War: Third World Interventions and the Making of Our Times* (Cambridge: Cambridge University Press, 2005).

44. The spread of river basin development and planning was epitomized by the internationalization of the experiences of the Tennessee Valley Authority and the "TVA model" from the 1930s through the 1970s, although to a large extent the model reflected a mythology regarding the TVA's supposed advancements and successes in river basin development. For an incisive analysis, see Ekbladh, *Great American Mission*, 77–113. For a more specific example, see Heather Hoag, "Transplanting the

TVA? International Contributions to Postwar River Development in Tanzania," *Comparative Technology Transfer and Society* 4 (2006): 247–68.

45. On this latter point, see Gabriel Kolko, *Confronting the Third World: United States Foreign Policy, 1945–1980* (New York: Pantheon Books, 1988); David Slater, "Reimagining the Geopolitics of Development: Continuing the Dialogue," *Transactions of the Institute of British Geographers* 19 (1994); and Zachary Karabell, *Architects of Intervention: The United States, the Third World, and the Cold War, 1946–1962* (Baton Rouge: Louisiana State University Press, 1999).

46. There is also the intriguing observation that certain technologies, large dams among them, transcend political ideologies. In the case of large dams, the architects of both Soviet and American economic development throughout the twentieth century perceived the harnessing of rivers as a critical means to promote national industrialization goals via the production of cheap electricity. And as Donald Worster has eloquently shown for the western United States, and Paul Josephson for the Soviet Union, the storage function of large dams was a key factor in the expansion of industrial agriculture. See Worster, *Rivers of Empire*; and Paul R. Josephson, *Industrialized Nature: Brute Force Technology and the Transformation of the Natural World* (Washington, DC: Island Press, 2002).

47. Geographers and related scholars working under the rubric of critical geopolitics have usefully explored Cold War geopolitics from cultural and political angles. See Joanne P. Sharp, *Condensing the Cold War: Reader's Digest and American Identity* (Minneapolis: University of Minnesota Press, 2001); and Neil Smith, *American Empire: Roosevelt's Geographer and the Prelude to Globalization* (Berkeley: University of California Press, 2004). However, the connections between US geopolitical strategies and the North-South dynamics of economic development during the Cold War era have received less attention within geography. See Klaus Dodds, *Global Geopolitics: A Critical Introduction* (Essex, UK: Prentice Hall, 2005), 55–60.

48. Slater, "Geopolitics of Development."

49. See Engerman, "Romance of Economic Development," and Westad, *Global Cold War*.

50. Gilman, *Mandarins*, 43.

51. Gearóid Ó Tuathail and John Agnew, "Geopolitics and Discourse: Practical Geopolitical Reasoning in American Foreign Policy," *Political Geography Quarterly* 11 (1992).

52. For ambitious theoretical efforts to stake out the conceptual terrain of a geopolitics of development, see Ó Tuathail and Agnew, "Geopolitics and Discourse"; and David Slater, *Geopolitics and the Post-Colonial: Rethinking North-South Relations* (Malden, MA: Blackwell, 2004).

53. See Scott, *Seeing like a State*.

54. John Agnew, *Hegemony: The New Shape of Global Power* (Philadelphia: Temple University Press, 2005), 1–2. However, the concept of hegemony as deployed originally by Gramsci and more recently by contemporary work in the social sciences also directs attention to the everyday processes whereby certain individuals and groups internalize thoughts and actions that may be inimical to their broader interests within society. See Antonio Gramsci, *Selections from the Prison Notebooks* (New York: International Publishers, 1971), 12–14. For the recent applications to water-society relations, see Michael Ekers and Alex Loftus, "The Power of Water: Developing Dialogues between Foucault and Gramsci," *Environment and Planning D: Society and Space* 26 (2008); and Trevor Birkenholz, "Groundwater Governmentality: Hegemony and Technologies of Resistance in Rajasthan's (India) Groundwater Governance," *Geographical Journal* 175 (2009).

55. Robert Cox, "Gramsci, Hegemony and International Relations: An Essay in Method," *Millennium: Journal of International Studies* 12 (1983): 171.

56. See Martin Müller, "Opening the Black Box of the Organization: Socio-Material Practices of Geopolitical Ordering," *Political Geography* 31 (2013): 379–88. Müller, in his cogent critique of critical geopolitics for its inattention to the actions of bureaucracies as highly influential geopolitical agents, rightly focuses on organizations (e.g., foreign ministries, think tanks, peace and conflict NGOs, etc.) whose mission has an explicit foreign policy aim or angle. By contrast, I examine the Bureau of Reclamation as a salient geopolitical agent, a largely unexpected and unasked-for role within an organization devoted to water resource development.

57. In his exegesis of the historical role of US foreign aid in the construction of the modern state of Egypt, Timothy Mitchell argues that what is missing from most accounts of technology transfer from the industrialized to the "developing" world is a failure to come to grips with, first, the inseparability of the human and nonhuman worlds and, second, the politics of technical expertise. See Timothy Mitchell, *Rule of Experts: Egypt, Techno-Politics, Modernity* (Berkeley: University of California Press, 2002). Gabrielle Hecht's definition of technopolitics as a "concept that captures the hybrid forms of power embedded in technological artifacts, systems, and practices" also resonates with my usage. See Gabrielle Hecht, "Introduction," in *Entangled Geographies: Empire and Technopolitics in the Global Cold War*, ed. Gabrielle Hecht (Cambridge, MA: MIT Press), 3.

58. Mitchell, *Rule of Experts*, 42–43.

59. For salient examples, see Langdon Winner, "Do Artifacts Have Politics?," *Daedalus* 109 (1980); Kristin Asdal, Brita Brenna, and Ingunn Moser, "Re-Inventing Politics of the State: Science and the Politics of Contestation," in *Technoscience: The Politics of Interventions*, ed. Kristin Asdal, Brita Brenna, and Ingunn Moser (Oslo: Oslo Academic Press, 2007), 7–53; and Dimitris Papadopoulos, "Alter-Ontologies: Towards a Constituent Politics in Technoscience," *Social Studies of Science* 41 (2011).

60. There are, of course, notable exceptions to the omissions highlighted here within STS and other disciplines. For example, drawing on a combination of STS approaches and political ecology, Phadke offers an enlightening analysis of a dam project in rural Maharashtra state (India) that served as a collective, and progressive, expression of political will on the part of diverse social groups. See Phadke, "Assessing Water Scarcity." Nega's superb account of a small dam in northern Minnesota (United States) adopts and extends an actor-network framework to draw attention to the ways in which technologies act within and are generative of political relations comprising humans and nonhumans. See Nega, "Saving Wild Rice." In general, however, dams and water resource development have received much greater attention within the history of technology literature, albeit with a bias toward US-based case studies. For a helpful overview of this literature, see Stine and Tarr, "Intersection of Histories." In terms of STS and Cold War geopolitics, Doel examines the growth of the earth/environmental sciences and related technologies as spurred by US military expenditures in response to needs understood in relation to Cold War objectives. See Ronald E. Doel, "Constituting the Postwar Earth Sciences: The Military's Influence on the Environmental Sciences in the USA after 1945," *Social Studies of Science* 33 (2003). For a similar exegesis of a variety of technological and scientific things and processes as co-produced with Cold War geopolitics, see Gabrielle Hecht, ed., *Entangled Geographies: Empire and Technopolitics in the Global Cold War* (Cambridge, MA: MIT Press, 2011).

61. Gearóid Ó Tuathail, "General Introduction: Thinking Critically about Geopolitics," in *The Geopolitics Reader*, 2nd ed., ed. Gearóid Ó Tuathail, Simon Dalby, and Paul Routledge (New York: Routledge, 2006), 11.

62. Tania Murray Li, *The Will to Improve: Governmentality, Development, and the Practice of Politics* (Durham, NC: Duke University Press, 2007), 7–8.

63. James Ferguson, *The Anti-Politics Machine: "Development," Depoliticization, and Bureaucratic Power in Lesotho* (Minneapolis: University of Minnesota Press, 1994), 17–21.

64. This era is indelibly captured in the context of the United States by Reisner's description of the "go-go years," a period of intense dam-building competition between the Bureau and the US Army Corps of Engineers that witnessed a remarkable increase in dammed rivers over the course of the 1950s and 1960s. See Marc Reisner, *Cadillac Desert: The American West and Its Disappearing Water* (New York: Viking, 1986), 145–68. As the argument in chapter 6 attests, we may be entering or already in a new "Golden Era" of dam construction.

65. See Michael C. Robinson, *Water for the West: The Bureau of Reclamation, 1902–1977* (Chicago: Public Works Historical Society, 1979), 55; Donald Worster, "The Hoover Dam: A Study in Domination," in *Social and Environmental Effects of Large Dams*, vol. 2, *Case Studies*, ed. Goldsmith and Hildyard, 17–24; and David B. Billington and Donald C. Jackson, *Big Dams of the New Deal Era: A Confluence of Engineering and Politics* (Norman: University of Oklahoma Press, 2006).

66. Several excellent recent works highlight the long-term ideological impacts of the TVA on the collective imaginations of non-US states. See Biggs, "Reclamation Nations"; Klingensmith, "One Valley"; Adas, *Dominance by Design*, 219–79; and Ekbladh, *Great American Mission*, 77–113.

67. There is no doubt that the New Deal "infused new life into the Reclamation program," and throughout the 1930s the Bureau of Reclamation's domestic activities were greatly expanded. The average yearly expenditure on Bureau projects from 1933 to 1940 was US$52 million, a fivefold increase from the US$8.9 million average in the years previous. During the 1934 fiscal year the Denver office's staff increased from 200 to 750 full-time employees. See Robinson, *Water for the West*, 55–57.

68. David Ekbladh illuminates how the notion of the TVA provided an almost perfect vehicle, mobilized under the Point Four program's aegis, for disseminating American influence abroad because it brought together elements of technological prowess, modernization, economic efficiency, and democratization within a cohesive package. Any nation with underdeveloped river basins saw the TVA as an almost instant means of moving "forward" toward economic modernization. See Ekbladh, *Great American Mission*, 77–113.

69. For an important examination of the often overlooked need for engineers to play the role of "negotiator" throughout the course of specific projects, see Martin Reuss, "Seeing Like an Engineer: Water Projects and the Mediation of the Incommensurable," *Technology and Culture* 49 (2008).

70. This observation resonates, albeit in a simplistic fashion, with Michel Serres's understanding of time as braided and nonlinear. For Serres, "time flows in a turbulent and chaotic manner; it percolates." See Michel Serres (with Bruno Latour), *Conversations on Science, Culture, and Time*, trans. Roxanne Lapidus (Ann Arbor: University of Michigan Press, 1995), 59. Dams interweave multiple temporalities (as shown in the examples), whether constituted by sediments that are halted and accumulate over time within their reservoirs, by the years it takes to obtain sufficient funding, by the profound changes they make in seasonal flow dynamics, or by a wide array of other processes.

71. For example, one of the fundamental choices for dam builders in the western United States during the early twentieth century was whether to construct the traditional gravity dam or to use an emerging technology based on multiple-arch dam design. So-called gravity dams are characterized by a very thick base (the part of the dam forming its foundation and "connection" to underlying earthen material), relying on their tremendous mass to oppose water pressure, and thus require large amounts of concrete, earth, or other building materials. Arch dams, by contrast, distribute water pressure to a series of buttresses (constructed on the dam's downstream face) or the walls of the river's canyon, thus requiring fewer materials and associated expenses. Ultimately, these technical decisions are shaped by political forces and subsequently influence socioecological changes. See David M. Intracaso, "The Politics of Technology: The 'Unpleasant Truth about Pleasant Dam,'" *Western Historical Quarterly* 26 (1995): 333–52.

72. Tsing's notion of "scale-making projects"—used to make sense of the complex politics of ecological and social transformation in Kalimantan (Indonesia) over recent decades—wherein scale "is the spatial dimensionality necessary for a particular kind of view" is an important inspiration for this perspective. See Anna Tsing, *Friction: An Ethnography of Global Connections* (Princeton, NJ: Princeton University Press, 2005), 57–58.

73. Atsushi Akera refers to this approach as an "ecology of knowledge," a view that endeavors to "depict the full range of historical entities" (e.g., the institutions, organizations, knowledge/disciplines, artifacts, and actors) when seeking to understand a particular phenomenon that brings together ecological, social, and technological elements. The approach also recognizes the "challenge of moving across different scales of analysis" when the entities in question—in my case, large dams and the Bureau of Reclamation—are constituted through expertise, capital, and ideologies that are disseminated through space and time. See Atsushi Akera, "Constructing a Representation for an Ecology of Knowledge: Methodological Advances in the Integration of Knowledge and Its Various Contexts," *Social Studies of Science* 37 (2007): 424–33.

74. Arif Dirlik argues that world history is "inevitably about 'world-making'" and that its core principle, while fraught with multiple interpretations, "displays an urge not only to overcome a nation-bound historiography that seems to be incapable of addressing possibly the most important phenomena of the contemporary world which transcend national and even regional boundaries, but also to do so by overcoming the Eurocentrism of past world histories." While skeptical of global histories, Dirlik nevertheless argues for their relevance and significance. See Arif Dirlik, "Confounding Metaphors, Inventions of the World: What Is World History For?," in *Writing World History, 1800–2000*, ed. Benedikt Stuchtey and Eckhardt Fuchs (London: Oxford University Press, 2003), 94–96.

75. An exemplar in this regard is Allen E. Isaacman and Barbara S. Isaacman, *Dams, Displacement, and the Delusion of Development: Cahora Bassa and Its Legacies in Mozambique* (Athens: Ohio University Press, 2013).

76. There is a spate of recent work examining the intersections among race, gender, and geopolitics as manifested within US foreign policy, some with direct relevance to the present work. For example, African officials visiting the TVA in the 1940s and 1950s were often confronted with the overt segregation policies of the Jim Crow South and denied seating in restaurants and the like, encounters that mortified the African guests and embarrassed TVA and State Department officials. See Robert Rook, "Race, Water, and Foreign Policy: The Tennessee Valley Authority's Global Agenda Meets 'Jim Crow,'" *Diplomatic History* 28 (2004): 55–81. For a broader overview, see Mary

Dudziak, *Cold War Civil Rights: Race and the Image of American Democracy* (Princeton, NJ: Princeton University Press, 2000).

77. Unfortunately, charting this influence lies outside the scope of this work. For an astute, historically rich exegesis in this vein, see Ruth Oldenziel, *Making Technology Masculine: Men, Women and Modern Machines in America, 1870–1945* (Amsterdam: Amsterdam University Press, 1999). For a broader examination of gender and geopolitics, see Jennifer Hyndman, "Mind the Gap: Bridging Feminist and Political Geography through Geopolitics," *Political Geography* 23 (2004): 307–22.

78. See Georg Iggers, *Historiography in the Twentieth Century: From Scientific Objectivity to the Postmodern Challenge* (Middletown, CT: Wesleyan University Press, 1997), 145.

79. See Hayden White, *The Content of the Form: Narrative Discourse and Historical Representation* (Baltimore: Johns Hopkins University Press, 1987), 164–65.

80. See Martin Müller, "Reconsidering the Concept of Discourse for the Field of Critical Geopolitics: Towards Discourse as Language and Practice," *Political Geography* 27 (2008): 323–24.

81. In selecting the cases that constitute the focal points of the chapters, I wanted to express both the geographical and temporal range of the Bureau's international engagements. The selection process was of course driven to some extent by the availability of material in the archival records of the Bureau and of the State Department. I chose to generate an appendix of additional Bureau experiences across the tricontinental world in order to preserve the richness of the many additional engagements that share similarities with, yet maintain crucial differences from, the cases featured in the chapters. Ultimately, I opted for a case study methodology because I, like others, perceive knowledge as context-dependent, and because the case study offers a comprehensive (albeit flawed, as is all knowledge production) means of coming to grips with the messiness of emergent phenomena that are at root hybrids of social, technological, and ecological processes. See Bent Flyvbjerg, "Five Misunderstandings about Case-Study Research," *Qualitative Inquiry* 12 (2006): 221–24.

82. Numerous case studies of specific projects—in a diversity of geographical settings—highlight how transfers of dam-related technologies to "developing" societies are related to political rationales at national and global levels. For an example from southern Africa, see Allen Isaacman and Chris Sneddon, "Toward a Social and Environmental History of the Building of Cahora Bassa Dam," *Journal of Southern African Studies* 26 (2000): 597–632. For other parts of Africa, see David Hart, *The Volta River Project: A Case Study in Politics and Technology* (Edinburgh: Edinburgh University Press, 1980); James McCann, "Ethiopia, Britain, and the Negotiations for the Lake Tana Dam, 1922–1935," *International Journal of African Historical Studies* 14 (1981): 667–69; and Hoag, "Transplanting the TVA?" In South and Central Asia, see Ben Crow, *Sharing the Ganges: The Politics and Technology of River Development* (Thousand Oaks, CA: Sage Publications, 1995); William Fisher, ed., *Toward Sustainable Development: Struggling over India's Narmada River* (London: M. E. Sharpe, 1995); and Nick Cullather, "Damming Afghanistan: Modernization in a Buffer State," *Journal of American History* 89 (2002): 512–37. And, finally, in Latin America, see Curtis Holder, "Contested Visions: Technology Transfer, Water Resources, and Social Capital in Chilascó, Guatemala," *Comparative Technology Transfer and Society* 4 (2006): 269–86.

83. My analysis in this regard is subjective and should be seen as incomplete. My labeling of Bureau engagements in specific places as "low," "moderate," or "high" is based on the following factors: length of Bureau visit; number of Bureau staff members in country; and types of investigations carried out (e.g., single schemes versus river

basin planning initiatives). This figure is read most usefully together with the entries in the appendix.

84. Robert J. Newell, *Reconnaissance Report: Nicaraguan Power Investigation Mission, Managua, Nicaragua* (Washington, DC: US Bureau of Reclamation, July 1951), i. Many of the reports, telegrams, communiqués, and meeting notes I quote throughout the book I have reproduced verbatim, even when language and grammar seem rough or poor, since I believe it is important to preserve the original tone of these resources.

85. Ibid., ii.

CHAPTER TWO

1. Arthur Powell Davis, Director of the Bureau of Reclamation, *1921 Fiscal Year Report*, Denver, Colorado, quoted in Michael Straus, letter to Howard P. Hughes, *Collier's*, June 12, 1952, RG 115, General Correspondence (1946–1960), box 9, Federal Records Center (hereafter FRC), NARA (Denver).

2. It would be a mistake to see this distribution of knowledge as a one-way affair. For several examples of cross-fertilization of water knowledge in the late nineteenth and early twentieth century between the primarily British engineers working in South Asia and their American counterparts in the western United States, see James Wescoat, "Wittfogel East and West: Changing Perspectives on Water Development in South Asia and the US, 1670–2000," in *Cultural Encounters with the Environment: Enduring and Evolving Geographic Themes*, ed. Alexander B. Murphy and Douglas L. Johnson (Lanham, MD: Rowman & Littlefield, 2000), 109–32.

3. Abel Wolman and W. H. Lyles, *John Lucian Savage, 1879–1967, Biographical Memoir* (Washington, DC: National Academy of Sciences, 1978), 229–30.

4. Agnew, *Hegemony*.

5. The report and its conclusions are recounted in George O. Pratt, *Foreign Activities of the Bureau of Reclamation* (for administrative use) (Denver: US Bureau of Reclamation, 1953), 2, RG 115, General Correspondence (1946–1960), box 4, FRC, NARA (Denver).

6. Many of the settlers in the western United States simply could not repay the fees required of them under specific Reclamation projects. See Robinson, *Water for the West*, 38–39.

7. Pratt, *Foreign Activities*, 3.

8. See S. Goldberg, "Ethiopia Enters New Era of Development," *Commerce Reports, February 9, 1931* (Washington, DC: Department of Commerce, 1931); and Science Service, "Giant Chain of Artificial Lakes Proposed by German to Irrigate Interior of Africa," *Washington Daily News*, October 8, 1936. For a close investigation of the Lake Tana project, see McCann, "Ethiopia, Britain, and the Negotiations," 667–69.

9. Pratt, *Foreign Activities*, 3.

10. See Smith, *American Empire*.

11. Agnew, *Hegemony*, 1–2.

12. Ibid., 3–4.

13. Robinson, *Water for the West*, 71.

14. See Adas, *Dominance by Design*.

15. Agnew, *Hegemony*, 52.

16. Ibid., 91–92.

17. For a comprehensive review of this turn toward agency in geopolitics scholarship, see Merje Kuus, "Professionals of Geopolitics: Agency in International Politics," *Geography Compass* 2 (2008). As noted in chapter 1, a more "strongly objective" study

would analyze the highly gendered nature of engineering in general and the masculinities that are performed through a variety of technological means. For a fuller explanation of "strong objectivity" and its application to nature-technology-society studies, see Sandra Harding, "After the Neutrality Debate: Science, Politics, and 'Strong Objectivity,'" *Social Research* 59 (1992).

18. Kuus, "Professionals," 2071.

19. This would, of course, become the now (in)famous Three Gorges Dam, completed by the Chinese government in 2003 and responsible for the displacement of over 1 million people and numerous socioecological disruptions. It is, as of last assessment, the largest human construction ever completed. On the Three Gorges Dam, see Baruch Boxer, "China's Three Gorges Dam: Questions and Prospects," *China Quarterly* 113 (1988): 94–108; Philip M. Fearnside, "China's Three Gorges Dam: 'Fatal' Project or Step towards Modernization?," *World Development* 16 (1988): 615–30; and Dai Qing, *Yangtze! Yangtze!*, ed. P. Adams and J. Thibodeau (London: Probe International, 1994). Throughout the 1950s the Bureau also conducted semiformal consulting work with the newly formed Taiwanese government after the Nationalists evacuated the mainland in 1949 (see the appendix). Somewhat ironically, the Bureau maintained linkages to the Communist Chinese government and consulted on the designs for the Three Gorges project in the 1980s and 1990s. See J. F. LaBounty, "Assessment of the Environmental Effects of Constructing the Three Gorge Project on the Yangtze River," *Water International* 9 (1984): 10–17.

20. Wolman and Lyles, *John Lucien Savage*, 227. The Minidoka project, completed in 1909, was the first project to reveal the potential of generating, distributing, and selling hydroelectricity as a key element in Bureau projects, particularly in terms of financing less economically attractive irrigation projects. See Robinson, *Water for the West*, 29.

21. Numerous studies have documented this strain of American environmentalism. For a compelling examination, see Clayton R. Koppes, "Efficiency, Equity, Esthetics: Shifting Themes in American Conservation," in *The Ends of the Earth: Perspectives on Modern Environmental History*, ed. Donald Worster (Cambridge: Cambridge University Press, 1988).

22. See Robinson, *Water for the West*, 69–71.

23. See Karin Ellison, "Explaining Hoover, Grand Coulee, and Shasta Dams: Institutional Stability and Professional Identity in the U.S. Bureau of Reclamation," in *Reclamation, Managing Water in the West: The Bureau of Reclamation: History Essays from the Centennial Symposium*, vol. 1, ed. U.S. Department of the Interior, Bureau of Reclamation (Denver: US Department of Interior, Bureau of Reclamation, 2008), 221–48. Bureau historian Michael Robinson described this period in the following terms: "The Bureau's organization was a cohesive, committed group dedicated to natural resource development for the benefit of all people. The agency manifested enormous esprit de corps and was alive with fresh, creative ideas." See Robinson, *Water for the West*, 71.

24. Wolman and Lyles, *John Lucien Savage*, 227.

25. In 1933 Savage and Bureau staff also designed the first two projects—the Norris and Wheeler Dams—on the Tennessee River, since the newly created Tennessee Valley Authority (TVA) had not had a chance to retain its own design engineer. See B. Rhodes, "From Cooksville to Chungking: The Dam-Designing Career of John L. Savage," *Wisconsin Magazine of History* 72 (1989): 261.

26. Wolman and Lyles, *John Lucien Savage*, 228–30.

27. For an excellent overview of the various forms this politicization took in the western United States, see Billington and Jackson, *Big Dams*. The Bureau's political activities

are also famously documented, albeit in a more polemic fashion, in Reisner, *Cadillac Desert*.

28. See Rhodes, "Cooksville," 260–61.

29. "The Dams that Jack Builds," *Newsweek*, April 2, 1945, 50.

30. Joseph Phillips, "The Best Dam Man in the Business," *Colorado Wonderland*, December 1953, 36, quoted in Wolman and Lyles, *John Lucien Savage*, 230.

31. Department of State Press Release No. 321, March 12, 1945, RG 115, Yangtze Gorge Project (1943–1949), box 9, NARA (Denver).

32. Savage's work and demeanor so impressed Kanwar Sain, one of India's key dam-building engineers in the mid-twentieth century, that he later described him as a karm yogi, a man "with a spiritual devotion to duty and righteousness, energetic and focused, yet dispassionate and above any thought of self-interest or gain." See Klingensmith, "One Valley," 229.

33. Telegram (890H.64A/28a), the Secretary of State to the Minister in Afghanistan (Engert), December 16, 1943, in *Foreign Relations of the United States* (hereafter FRUS) *Diplomatic Papers 1943*, vol. 4, *The Near East and Africa* (Washington, DC: GPO, 1943), 63.

34. H. W. Bashore, letter to Haldore Hanson (Department of State), December 4, 1943, RG 115, Yangtze Gorge Project (1943–1949), box 3, NARA (Denver).

35. G. H. Shaw, letter to John L. Savage, December 13, 1943, RG 115, Yangtze Gorge Project (1943–1949), box 3, NARA (Denver).

36. It should also be noted that the Nationalist Republican government in China had a long history of employing foreign experts and bending their advice to fit its own political and developmental agendas. See Margherita Zanasi, "Exporting Development: The League of Nations and Republican China," *Comparative Studies in Society and History* 49 (2007).

37. A partial list includes Herbert Feis, *The China Tangle: The American Effort in China from Pearl Harbor to the Marshall Mission* (Princeton, NJ: Princeton University Press, 1953); Tang Tsou, *America's Failure in China, 1941–50* (Chicago: University of Chicago Press, 1963); and Michael Schaller, *The U.S. Crusade in China, 1938–1945* (New York: Columbia University Press, 1979). Schaller's work is particularly rich in archival detail and analysis covering the period roughly coinciding with Savage's time in China.

38. This regime is described vividly in Michael Schaller, *The United States and China in the Twentieth Century*, 2nd ed. (New York: Oxford University Press, 1990), 53–86.

39. A fuller account of the long and complex political-economic and cultural relationship between the United States and China is beyond the scope of the present work. Schaller, for example, details the "myth of the China market" that captured the imagination of both the US government and public in the 1920s and 1930s. The missionary impulse that had characterized previous attitudes toward China in the popular American imagination in earlier decades was coupled to this market myth, which "conjured up an image of 450 million consumers able to absorb vast amounts of European and American exports." See Schaller, *United States and China*, 18–19. During the period of discussions over the Yangtze Gorge Dam, State Department and US embassy communications consistently made reference to the vast economic potential of China that would almost certainly be unlocked through resource development, all to America's advantage in the long run.

40. Quoted in Schaller, *United States and China*, 53.

41. Ibid., 53–58.

42. William E. Warne, *The Bureau of Reclamation* (New York: Praeger, 1973), 229–30. Boxer argues that Savage's role as adviser to the NRC and his personal connections to

Soong and Wenhao, both of whom were intimately involved in the planning of the Yangtze Gorge project, were critical in implanting and solidifying the idea of a dam in the Three Gorges region, an "ideal setting" in the consciousness of generations of Chinese engineers. See Boxer, "China's Three Gorges Dam," 101.

43. See Deirdre Chetham, *Before the Deluge: The Vanishing World of the Yangtze's Three Gorges* (New York: Palgrave Macmillan, 2002), 2–4.

44. J. L. Savage, "Excerpts from Preliminary Report on Yangtze Gorge Project, Chungking, China," November 9, 1944, RG 115, Yangtze Gorge Project (1943–1949), box 10, NARA (Denver).

45. "The Dams that Jack Builds," 50.

46. Memorandum for Assistant Secretary Straus, "Chronology of Events Relating to the Proposed Agreement between the Bureau of Reclamation and the National Resources Commission of China for Preparation of the Yangtze Gorge and Tributary Reports," June 6, 1945, 1–2, RG 115, Yangtze Gorge Project (1943–1949), box 9, NARA (Denver).

47. Ibid., 2.

48. Ibid., 3.

49. Ibid.

50. Joseph C. Grew, letter to Secretary of Interior Harold Ickes, May 18, 1945, RG 115, Yangtze Gorge Project (1943–1949), box 10, NARA (Denver).

51. Michael Straus, Memorandum for the Files re: "Feasibility of Action by the Bureau of Reclamation in Performance of Services for the Chinese Government in the Development of Irrigation and Electrical Energy in China," June 9, 1945, RG 115, Yangtze Gorge Project (1943–1949), box 10, NARA (Denver).

52. Harold L. Ickes, letter to Department of State, Acting Secretary Grew, April 25, 1945, RG 115, Yangtze Gorge Project (1943–1949), box 9, NARA (Denver).

53. Ibid.

54. Michael Straus, Memorandum for Secretary Chapman, May 21, 1945, RG 115, Yangtze Gorge Project (1943–1949), box 9, NARA (Denver).

55. John L. Savage, letter to Michael W. Straus, assistant secretary, Department of the Interior, June 27, 1945, RG 115, Yangtze Gorge Project (1943–1949), box 10, NARA (Denver).

56. "Notes on a Meeting of General Marshall with the President, Mr. Byrnes, and Admiral Leahy at 3:30 p.m.," December 11, 1945; in FRUS *Diplomatic Papers 1945*, vol. 7, *The Far East, China* (Washington, DC: GPO, 1945), 745–47.

57. James R. Shepley, Memorandum to General Marshall, December 19, 1945; in *FRUS Diplomatic Papers 1945*, vol. 7, 774–77.

58. Patrick J. Hurley, Memorandum to the Secretary of State, August 8, 1945; in *FRUS Diplomatic Papers, 1945*, vol. 7, 1434–35.

59. Rhodes, "Cooksville," 266–70.

60. Alexander Schnee, Memorandum of Conversation with Mr. Corfitzen (Bureau of Reclamation) re: "Contract between the Bureau of Reclamation and the National Resources Commission of China for the Designing of the Yangtze Dam," May 29, 1947, RG 115, Yangtze Gorge Project (1943–1949), box 21, NARA (Denver).

61. William C. Jones and Marsha Freeman, "Three Gorges Dam: The TVA on the Yangtze River," *21st Century* (Fall 2000), 30. Available at http://www.21stcenturysciencetech.com/articles/Fall_2000/TVA_Yangtze.pdf.

62. See Boxer, "China's Three Gorges Dam," 101.

63. Michael Straus, *26,000 Miles along Reclamation Street: Report on South Asian Food, Water, and Power Development* (Washington, DC: Department of the Interior, February and January 1951).

64. Michael Straus, *Why Not Survive?* (New York: Simon and Schuster, 1955).

65. Alfred R. Golze, "Technical Assistance to Foreign Countries by the Bureau of Reclamation" (Department of the Interior, Information Service, October 22, 1959), 1–6, RG 115, General Correspondence (1946–1960), box 4, FRC, NARA (Denver). It should be noted that over half of the nations receiving Bureau technical assistance were part of a training program and did not necessarily host Bureau experts.

66. Gilbert G. Stamm, "Bureau of Reclamation International Technical Assistance in Development of Arid Lands" (paper presented at the International Conference on Arid Lands in a Changing World, American Association for the Advancement of Science, Tucson, AZ, June 13, 1969), 1. Over the course of the 1950s and 1960s, the Bureau's role in technical assistance to Cold War allies and would-be allies of the United States assumed a variety of forms. Staff assignments ranged from multiyear tasks involving entire teams of engineers and related experts to those involving a single engineer for several weeks. Accordingly, the actual tasks performed by Bureau employees included an assortment of activities associated with water resource development. In some cases, Bureau experts were asked to simply consult on an individual project's design or to evaluate a specific aspect of project construction or operation. In other cases, such as that described in chapter 3 in Lebanon, Bureau staff participated in water resource development activities at a much deeper level.

67. Pratt, *Foreign Activities*.

68. Arthur M. Schlesinger Jr., *The Vital Center: The Politics of Freedom* (Boston: Houghton Mifflin, 1949), 233.

69. Adas, *Dominance by Design*, 6.

70. For a well-thought-out argument in this vein, see Paul R. Josephson, "'Projects of the Century' in Soviet History: Large-Scale Technologies from Lenin to Gorbachev," *Technology and Culture* 36 (1995).

71. See Gilbert Rist, *The History of Development: From Western Origins to Global Faith* (London: Zed Books, 2002); and Mark Berger, *The Battle for Asia: From Decolonization to Globalization* (London: RoutledgeCurzon, 2004).

72. As Michael Adas notes, the "high-tech TVA model of development assistance would have a profound impact on U.S. foreign policy after the war." See Adas, *Dominance by Design*, 208–29. Ironically, the architects of this foreign policy paid little attention to the growing number of critics of the TVA as it was actually evolving (as opposed to its idealized grand vision). See chapter 3 for an expanded discussion of this critique.

73. As Mitchell argues, this lack of effectiveness is a direct result of the messiness and hybridity of technological, political, and environmental processes brought into contact under the rubric of development. The original intent of any development initiative, whether a mosquito eradication program or a large dam, is overwhelmed and often made irrelevant by the ways in which the material world shifts "in and out of human forms, or occurs as arrangements . . . that are social as well as natural, technical as well as material." See Mitchell, *Rule of Experts*, 52.

74. As noted in chapter 1, this is a central idea expressed by James Ferguson in his interpretation of development as practiced in the twentieth century. The powerful interests "at work" in a development initiative "can only operate through a complex set of social and cultural structures so deeply embedded and so ill-perceived that the outcome may be only a baroque and unrecognizable outcome of the original intention." See Ferguson, *Anti-Politics Machine*, 17.

75. Wolman and Lyles, *John Lucien Savage*, 225. There is an obvious need, as mentioned in chapter 1, to undertake analyses of the highly gendered character of the Bureau and of the engineering profession more generally.

76. Adas, *Dominance by Design*, 226–27.

CHAPTER THREE

1. Clapp's quotations are taken from "U.N. Plans Arab TVA to Provide Refugees Work," *Chicago Daily Tribune*, December 13, 1949.

2. Straus, *26,000 Miles*, foreword.

3. See Schlesinger, *Vital Center*, 233.

4. Tsing, *Friction*, 58.

5. As noted in chapter 1, I recognize that effective case narratives have a certain "irreducible quality" that makes them difficult to summarize neatly or completely and reflects the messiness of material existence. Throughout the book, I strive to leave sufficient room in the case study for readers to develop their own conclusions. See Flyvbjerg, "Five Misunderstandings," 237.

6. Molle, "River-Basin Planning," 492.

7. See Worster, *Rivers of Empire*.

8. Quote by French engineer Aimé Thomé de Gamond in 1871, cited in Molle, "River-Basin Planning," 5.

9. For histories of the idea and practice of river basin thinking in the United States, see Worster, *Rivers of Empire*; and Wescoat, "'Watersheds' in Regional Planning." For the evolution of the idea in Europe, see Molle, "River-Basin Planning"; and Eric Swyngedouw, "Technonatural Revolutions: The Scalar Politics of Franco's Hydrosocial Dream for Spain, 1939–1975," *Transactions of the Institute of British Geographers* 32 (2007): 9–28.

10. See Gordon Clapp, *The TVA: An Approach to the Development of a Region* (Chicago: University of Chicago Press, 1955).

11. Only much later, roughly in the early 1990s, did the idea of integrated water resource management (IWRM) and its emphasis on the maintenance of flows for environmental goals become part of basin thinking. For a general history of the concept, see Muhammad Mizanur Rahman and Olli Varis, "Integrated Water Resources Management: Evolution, Prospects and Future Challenges," *Sustainability: Science, Practice, & Policy* 1 (2005): 15–21.

12. The idea of the TVA and its extension to various parts of the world as a component of American efforts at modernizing the world's "less advanced" regions has been aptly delineated in several recent works. For a general overview, see Adas, *Dominance by Design*, 219–79. For application of the TVA "dream" in India, see Klingensmith, "One Valley." For the ambitious plan termed a "TVA on the Mekong" by Lyndon Johnson, see Ekbladh, *Great American Mission*, 190–225; elements of this story are covered in chapter 5.

13. See David Slater, "Geopolitical Imaginations across the North-South Divide: Issues of Difference, Development and Power," *Political Geography* 16 (1997): 641.

14. Latham, *Modernization as Ideology*, 3–4.

15. David Ekbladh, "'Mr. TVA': Grass-Roots Development, David Lilienthal, and the Rise and Fall of the Tennessee Valley Authority as a Symbol for U.S. Overseas Development, 1933–1973," *Diplomatic History* 26 (2002): 338–39.

16. Arguably the most influential of these advocates was former TVA Chairman David Lilienthal, who upon his resignation from the organization after World War II relent-

lessly championed the TVA's capacity to simultaneously spur economic development and promote social democracy. See Ekbladh, "'Mr. TVA.'"

17. Klingensmith, "One Valley," 41.

18. Ibid., 43–45.

19. See Karen M. O'Neill, "Why the TVA Remains Unique: Interest Groups and the Defeat of New Deal River Planning," *Rural Sociology* 67 (2002): 163–82.

20. In the 1940s this same coalition decried river basin planning initiatives as a dangerous expansion of federal power. See Brian Black, "Organic Planning: The Intersection of Nature and Economic Planning in the Early Tennessee Valley Authority," *Journal of Environmental Policy and Planning* 4 (2002): 157–68. By the 1950s President Eisenhower was referencing the TVA as an example of "creeping socialism," revealing the extent to which the project had become the focus of ideological struggle during that era. See John Oliver, "The TVA Power Program," *Current History* 34 (1958): 264. As the decades have passed, the TVA has also been challenged on the grounds of its effectiveness, with many critics noting that its accomplishments have been "neither as original nor as inclusive as its boosters asserted" (see Ekbladh, "'Mr. TVA,'" 344) and that its social and ecological disruptions have been substantial. On this latter point, see William U. Chandler, *The Myth of the TVA: Conservation and Development in the Tennessee Valley, 1933–1980* (Cambridge, MA: Ballinger, 1984). More recently, the TVA has focused its operations almost exclusively on the generation and distribution of electricity. See O'Neill, "Why the TVA."

21. Adas, *Dominance by Design*, 208–9.

22. A host of countries around the world adopted TVA-style river basin development approaches in the 1950s and 1960s. Some of the most prominent examples include the Damodar Valley in India, the Gal Oya basin in Sri Lanka, the São Francisco basin in Brazil, the Helmand Valley in Afghanistan and the Magdalena-Cauca basin in Colombia. See Christopher J. Barrow, "River Basin Development Planning and Management: A Critical Review," *World Development* 26 (1998): 175–76.

23. See William Rowley, *The Bureau of Reclamation: Origins and Growth to 1945*, vol. 1 (Denver: Bureau of Reclamation, 2006), 18.

24. In the United States, the early twentieth-century regime of Theodore Roosevelt promoted multipurpose resource use projects and holistic planning for river basins as important conservationist tenets, and the creation of the Reclamation Program in 1902 incorporated these principles into its water resource philosophy. See Robinson, *Water for the West*, 25–27. Marc Reisner, citing Robinson's authorized history, refers to the large hydroelectric projects on the Colorado River as "cash register dams" due to their use within a murky kind of river basin "accounting" practiced by the Bureau that made financially questionable irrigation schemes look justifiable when seen as part of an overall basin program. See Reisner, *Cadillac Desert*, 136. Similarly, a former assistant commissioner of reclamation called the power from hydroelectric dams the "paying partner" of irrigation. See Warne, *Bureau of Reclamation*, 86.

25. On the failure of a basin-wide governance authority in the Columbia basin, see Billington and Jackson, *Big Dams*, 189–92.

26. Harry S. Truman, "Inaugural Address," January 20, 1949, http://www.trumanlibrary.org/whistlestop/50yr_archive/inagural20jan1949.htm. The Point Four program of aid and technical assistance has been commonly cited as the beginning of the global "development era," within which the United States was, according to President Truman, "to aid the efforts of the peoples of economically underdeveloped areas to de-

velop their resources and improve their working and living conditions." See Truman, "Inaugural Address." Most of the Bureau's early foreign operations were funded and directed in-country under the auspices of the Point Four administrative apparatus located within the US State Department. From its initiation, it was clear that the Point Four program reflected the economic and geopolitical interests of the United States. See Samuel P. Hayes, "Point Four in United States Foreign Policy," *Annals of the American Academy of Political and Social Science* 268 (1950): 27–35. Point Four was also deeply linked to American ideas of modernization and the United States' "unique" position within world politics in the years following World War II. Economic and technical assistance to "underdeveloped" regions was conceived from the start as an important component of foreign policy. For details on Point Four and its ideological and institutional origins, see Ekbladh, "'Mr. TVA,'" 77–113; and Latham, *Modernization as Ideology*, 70–71.

27. Harry S. Truman, "Remarks at the Women's National Democratic Club Dinner, November 8, 1949," http://www.trumanlibrary.org/publicpapers/.

28. Assistant Secretary, Memorandum re: Foreign activities—Implementation of "Point Four" program of technical cooperation, April 1, 1949, RG 115, General Correspondence (1946–1960), box 4, FRC, NARA (Denver). William Warne, former assistant commissioner of reclamation, eventually assumed leadership of the Point Four technical assistance program in Iran, where he became centrally involved with the Iranian government's water resource development programs in the 1950s. Not surprisingly, these programs were confounded by internal bureaucratic struggles and tensions over development priorities. For his account, see William Warne, *Mission for Peace: Point 4 in Iran* (Indianapolis: Bobbs-Merrill, 1956).

29. See Klingensmith, "One Valley." Other research has also shown a high correlation between the geographical focus of the Bureau's training programs for foreign nationals (i.e., the trainees' countries of origin) during the Cold War period and those regions and countries of high geopolitical interest to the United States. James Wescoat, Roger Smith, and David Schaad, "Visits to the U.S. Bureau of Reclamation from South Asia and the Middle East, 1946–1990: An Indicator of Changing International Program and Politics," *Irrigation and Drainage Systems* 6 (1992): 55–67.

30. Chief Engineer, Memorandum to the Commissioner re: Expanded Participation under "Point 4," April 15, 1949, RG 115, General Correspondence (1946–1960), box 3, FRC, NARA (Denver).

31. Robinson, *Water for the West*, 56–61.

32. L. W. Damours, "Memorandum to Commissioner re: Foreign activities—A Review," July 7, 1965, RG 115, General Correspondence (1944–1990), box 8, NARA (Denver).

33. Chief Engineer, Memorandum to the Commissioner.

34. Michael Straus, Memorandum to Chief Engineer, Denver, re: Foreign Activities—Performance of Engineering Work for Foreign Governments, April 18, 1950, RG 115, General Correspondence (1946–1960), box 3, FRC, NARA (Denver).

35. In 1948 the United States Congress passed the Information and Educational Exchange Act (the Smith-Mundt Act), which, while primarily concerned with regulating the use of propaganda both overseas and domestically, also provided legal authority to the Bureau and similar technical bureaucracies to provide staff for international activities. The Smith-Mundt Act gave the federal government the authority to fund the training of foreign nationals, to send personnel overseas, and to provide funding for technical assistance overseas. See Brit Allan Storey, "The Bureau of Reclamation and International

Water Development" (paper presented at Water in History: Global Perspectives on Politics, Economy, and Culture, University of Wales, Aberystwyth, Wales, July 8–11, 1999), 30–31.

36. This point is underscored by the authorized histories of the Bureau penned by William Warne (see note 28) and Michael Robinson. Warne speaks of the Bureau as a "high-morale outfit" whose staffers are justifiably proud of participating in "one of the most exciting phases of the winning of the West" in the United States, and who "wear the old school tie of comradeship in reclamation." Warne, *Bureau of Reclamation*, v–vi. Robinson quotes approvingly an *Engineering News-Record* editor who was "profoundly stirred" by the Bureau people "who take great pride in the work they are doing" and "display a personal responsibility for maintaining and enhancing the prestige of [their] organization." Robinson, *Water for the West*, 71.

37. Pratt, *Foreign Activities*, 9.

38. G. W. Lineweaver, Memorandum to Secretary of the Interior re: Brief Inspection Tour of Latin America by Bureau of Reclamation Group, April 3, 1951, RG 115, General Correspondence (1946–1960), box 12, FRC, NARA (Denver).

39. G. W. Lineweaver, Memorandum to Secretary of the Interior.

40. J. W. Dixon, Memorandum to George Pratt re: Hydrologic Aspects of Foreign Projects, July 21, 1952, RG 115, General Correspondence (1946–1960), box 12, FRC, NARA (Denver).

41. See Worster, *Rivers of Empire*; Donald C. Jackson, "Engineering in the Progressive Era: A New Look at Frederick Haynes Newell and the U.S. Reclamation Service," *Technology and Culture* 34 (1993): 539–74; and Billington and Jackson, *Big Dams*. These and other studies validate the claim that there is "no strict scientific logic to the advancement of technology; political considerations can arrest the progress of, or further the development of, any given project." Intracaso, "Politics of Technology," 348.

42. Quoted in Pratt, *Foreign Activities*, 10.

43. The International Bank for Reconstruction and Development (IBRD) and the International Development Association (IDA) are collectively known as the World Bank. Before creation of the IDA in 1960 as a means to offer concessional loans and grants to the poorest countries, the IBRD and the World Bank were basically synonymous. By the late 1960s "World Bank" had become the most common title used by the international development community, academics, and public media. I refer to "IBRD" in discussing the time period when that name was in frequent use and to the "World Bank" otherwise.

44. Feliks Bochinski and William Diamond, "TVA's in the Middle East," *Middle East Journal* 4 (1950): 55. The article identifies the Nile, Orontes (Syria), Yarmouk-Jordan, Litani (Lebanon), Karun (Iran), and Tigris-Euphrates basins—as well as several valleys in Turkey—as especially amenable to TVA-style programs.

45. See James Hudson, "The Litani River of Lebanon: An Example of Middle Eastern Water Development," *Middle East Journal* 25 (1971): 1–2.

46. US Bureau of Reclamation, *Reconnaissance Report on the Litani River Project, Lebanon* (Washington, DC: Department of the Interior, US Bureau of Reclamation, June 1951), 9.

47. Hudson, "Litani River," 1.

48. US Bureau of Reclamation, *Litani River Project*.

49. Michael Straus, letter to Arthur Goldschmidt re: Litani River Project, October 24, 1950, RG 115, General Correspondence (1946–1960), box 11, FRC, NARA (Denver).

50. US Bureau of Reclamation, *Litani River Project*, 5.

51. Robert F. Herdman, Telegram to Commissioner, Bureau of Reclamation, "Foreign Activities—Periodic Report on Status of Investigations—Litani Project, Lebanon," April 28, 1951, RG 115, Technical Correspondence, Agreements and Reports (1944–1990), box 8, FRC, NARA (Denver).

52. M. R. Lewis, "Outline of Discussion Held at American University of Beirut between Professors of the University and Members of the Litani River Commission," April 27, 1951, RG 115, International Reports (1914–1987), box 6, NARA (Denver).

53. Ibid.

54. Ibid.

55. For succeeding American regimes, and for the Eisenhower administration in particular, "the implantation of a European state in the Middle East was . . . seen as a way of exporting civilization and democracy to the region"; it was hoped that "the Zionist state could, over time, act in conjunction with the United States in fighting to keep Soviet influence out of the Middle East." See Westad, *Global Cold War*, 127–28.

56. "Department of State Position Paper," May 5, 1953; in *FRUS, 1952–1954*, vol. 9, part 1, *The Near and Middle East* (Washington, DC: GPO, 1953), 1210–11.

57. Ibid., 1210.

58. Clive Schofield, "Elusive Security: The Military and Political Geography of South Lebanon," *GeoJournal* 31 (1993): 149–61. For a detailed rendering of the Eisenhower administration's approach to Lebanon during this period, see Douglas Little, "His Finest Hour? Eisenhower, Lebanon, and the 1958 Middle East Crisis," in *Empire and Revolution: The United States and the Third World since 1945*, ed. Peter L. Hahn and Mary Ann Heiss (Columbus: Ohio State University Press, 2001), 17–47.

59. For a theoretically nuanced account of the role that various representations of the water of the Jordan River played in the early formation of the Israeli state, see Samer Alatout, "'States' of Scarcity: Water, Space, and Identity Politics in Israel, 1948–1959," *Environment and Planning D: Society and Space* 26 (2008): 959–82. Alatout deftly addresses the critical application of US political leverage toward use of the Jordan River and its social implications.

60. J. Jernegan, Memorandum by the Deputy Assistant Secretary of State for Near Eastern, South Asian, and African Affairs to the Secretary of State, August 10, 1953; in FRUS, 1952–1954, vol. 9, part 1, 1269–75.

61. A. Gardiner, "Memorandum of Conversation by the Politico-Economic Adviser in the Bureau of Near Eastern, South Asian, and African Affairs [New York]," June 2, 1954; in *FRUS, 1952–1954*, vol. 9, part 1, 1567–71.

62. See Alatout, "'States' of Scarcity."

63. Jernegan, Memorandum, 1271.

64. Michael R. Adamson, "'The Most Important Single Aspect of Our Foreign Policy'? The Eisenhower Administration, Foreign Aid, and the Third World," in *The Eisenhower Administration, the Third World, and the Globalization of the Cold War*, ed. Kathryn C. Statler and Andrew L. Johns (Lanham, MD: Rowman & Littlefield, 2006), 47.

65. "Department of State Position Paper," May 5, 1953; in *FRUS, 1952–1954*, vol. 9, part 1, 1214.

66. H. Peter, "Memorandum of Conversation by the Country Director of Technical Cooperation Service, Lebanon (Peter)," May 17, 1953; in *FRUS, 1952–1954*, vol. 9, part 1, 85–87.

67. Ibid.

68. J. Lobenstine, "Communication to Department of State," October 2, 1952; in *FRUS, 1952–1954*, vol. 9, part 1, 1014–16.

69. Bureau of Near Eastern, South Asian, and African Affairs, "Suggested Main Points of Approach toward Israel-Arab Settlement," January 14, 1955; in *FRUS, 1955–1957*, vol. 14, *Arab-Israeli Dispute, 1955* (Washington, DC: GPO, 1955), 9–19.

70. F. A. Houck, Memorandum to Bureau of Reclamation (Washington, DC) re: "Lower Litani River Investigations," September 8, 1952, RG 115, International Reports (1914–1987), box 6, NARA (Denver).

71. R. Kaser, Memorandum to S. B. Foster (Bureau of Reclamation, Washington, DC) re: "Selection of Overall Scheme of Development for Litani River Project, March 23, 1953," RG 115, International Reports (1914–1987), box 6, NARA (Denver).

72. US Bureau of Reclamation, *Development Plan for the Litani River Basin, Republic of Lebanon*, vol. 1, *General Description and Economic Analysis* (Beirut: Litani River Investigation Staff, 1954), i–ii.

73. Ibid., ii–iii.

74. Ibid., 32.

75. "Lebanon Plans Project without Foreign Aid," *Christian Science Monitor*, August 30, 1954, 4.

76. Originally, an additional impoundment (the Khardale Dam) was to be constructed in the lower portion of the basin to provide irrigation coverage (roughly 30,000 hectares) to southern Lebanon. See US Bureau of Reclamation, *Development Plan*, i–iii. The Khardale project was never executed due to political tensions. See Fady Comair, "Litani Water Management: Prospect for the Future" (paper presented at the Congres International de Kaslik–Liban, June 18–20, 1998), 3, http://www.samana.funredes .org/agua/files/geopolitique/COMIER.rtf.

77. "Flowing Sand Plugs Two Mile of Tunnel—and It Won't Come Out," *Engineering News-Record*, August 25, 1960, 40.

78. Underscoring the ways in which Cold War geopolitics could affect even modest river basin development schemes, the Soviet ambassador to Lebanon offered to finance the entire project in October 1955, although Lebanon had up to that point never accepted Soviet assistance. See "Israel Accuses Soviets," *Washington Post and Times Herald*, October 20, 1955. In general, the Soviets perceived Lebanon—and at this time, much of the Middle East, with the major exception of Egypt—as a relatively minor part of their broader "Third World" strategy, which in the late 1950s was reoriented toward building better relations with China. See Westad, *Global Cold War*, 65–72.

79. See Nimrod Raphaeli, "Development Planning: Lebanon," *Western Political Quarterly* 20 (1967): 714–28.

80. "Flowing Sand," 38–40.

81. Ibid., 38.

82. Raphaeli, "Development Planning," 717–19.

83. See Hudson, "Litani River." Conflicts over the financing of the project also delayed its completion. In 1963 the World Bank balked at payments on its $US27.4 million loan commitment due to what it considered unnecessary delays by the Lebanese government in meeting its funding obligations. See L. C. Tihany, letter to Department of State re: "Economic Development: Progress Report on the Litani Project," April 24, 1963, RG 115, Technical Correspondence, Agreements and Reports (1944–1990), box 26, FRC, NARA (Denver).

84. As early as 1950, Commissioner Michael Straus had voiced his hope that that a "substantial portion of the work" on the Litani project should be "delegated to private industry," preferably of American origin. See Straus, Memorandum to Chief Engineer.

85. See Kolko, *Confronting the Third World*, 89.

CHAPTER FOUR

1. Ryszard Kapuściński, *The Emperor: Downfall of an Autocrat* (New York: Vintage Books, 1983), 130–31.
2. Ashok Swain, "Ethiopia, the Sudan, and Egypt: The Nile River Dispute," *Journal of Modern African Studies* 35 (1997): 680–81.
3. Pratt, *Foreign Activities*, 8.
4. The Bureau and State Department archives present a constant challenge in terms of the "correct" names of countries, rivers, and specific dam projects. Many of the Bureau's overseas forays into the tricontinental world predate national independence and are thus categorized using colonial nomenclature (e.g., Northern Rhodesia, Ceylon). Moreover, the Bureau and State Department used no standard means of Anglicizing the names of rivers and dam sites from their original monikers in Chinese, Arabic, Amhara, Thai, or other languages.
5. See Wescoat, Smith, and Schaad, "Visits to the U.S. Bureau of Reclamation," 2. This number would certainly be much higher—by my estimation, on the order of ten thousand trainees—if the regions of Southeast and East Asia, Africa, and Latin America were to be included. The Bureau's archival records are remarkably incomplete on this topic.
6. Pratt, *Foreign Activities*, 9.
7. Kolko argues that America's primary interest in Third World states and societies during the twentieth century converged around the vital resources (e.g., mineral, plant-based) located within these regions and that US foreign policy must be understood within this crucial political economic context. Similarly, Westad sees political-economic objectives as equal in importance to ideological aims in constructing US and Soviet attitudes toward the Third World during this period. See Kolko, *Confronting the Third World*, 117–20; and Westad, *Global Cold War*, 152–57.
8. Pratt, *Foreign Activities*, 8.
9. Golze, "Technical Assistance," 8. In 1951, at the behest of Egyptian businessmen, both Westinghouse and American Cyanamid expressed strong interest in constructing the Aswan High Dam on the Nile River. See Laron, *Origins of the Suez Crisis*, 51.
10. The Bureau's annual budget fell from US$364 million in 1950 to US$165 million in 1955, no doubt curtailing the agency's overseas programs to some degree. See Robinson, *Water for the West*, 79–81. These budget challenges also explain the Bureau's nearly fanatical concern over securing reliable sources of funding from the State Department for its foreign ventures during the 1950s.
11. Floyd Dominy, "The U.S. Government: Its Engineering Role Abroad" (speech before the Second Institute for International Engineering, University of Colorado, Boulder, CO, September 24, 1964), RG 115, General Correspondence (1944–1990), box 8, NARA (Denver).
12. Ibid., 2.
13. Ibid., 3.
14. Division of Foreign Activities (Bureau of Reclamation), "Items for Consideration Relative to Possible Criticism of Reclamation's Overseas Work," attachment to Annual Report 1963, RG 115, General Correspondence (1944–1990), box 8, NARA (Denver).
15. Dominy, "U.S. Government," 3.
16. Division of Foreign Activities, "Items for Consideration."
17. L. W. Damours, "Memorandum to Commissioner re: Foreign Activities—A Review," July 7, 1965, RG 115, General Correspondence (1944–1990), box 8, NARA (Denver).
18. Given its dam- and Bureau-centric orientation, this chapter is necessarily inattentive to the complex political, economic, and cultural forces that defined the evolution

of the Ethiopian state and the Ethiopian people during the course of the twentieth century, particularly the period leading up to the revolutionary overthrow of Haile Selassie in 1974. I encourage consultations of the outstanding studies that do this history justice. See, for example, Bahru Zewde, *A History of Modern Ethiopia, 1855–1991*, 2nd ed. (Athens: Ohio University Press, 2001); Harold G. Marcus, *A History of Ethiopia*, 2nd ed. (Berkeley: University of California Press, 2002), 147–63; and John Markakis, *Ethiopia: The Last Two Frontiers* (Rochester, NY: James Currey, 2011), 108–30.

19. "The Dam: Its Power and Potentialities," *Ethiopia Observer* 4 (1960): 251.

20. The original agreement between the imperial Ethiopian government (IEG) and the United States was even more comprehensive, providing for studies of river basins in Eritrea and other basins within Ethiopian territory. During this period the Bureau team carried out water resource investigations on the Haddas and Ghinda Rivers in the Eritrea region. These surveys were based on work previously carried out by Italian engineers during the occupation period. See Thomas A. Clark, "Summary of Activities, July 1 to December 31, 1953," 1–3, RG 115, International Reports (1914–1987), box 4, NARA (Denver).

21. Marcus Gordon, "U.S. Technical Assistance Service for Ethiopia," letter to Lidj Yilma Deressa, Minister of Commerce and Industry, February 25, 1953, RG 115, International Reports (1914–1987), box 4, NARA (Denver).

22. Also found as "Coca" in numerous documents and literature on the Awash surveys.

23. "The Dam," 249.

24. Ibid.

25. Marcus Gordon, Memorandum to Department of State re: "Termination of Survey Work at Coca Dam Site During Rainy Season," August 12, 1953, RG 115, International Reports (1914–1987), box 4, NARA (Denver).

26. Thomas A. Clark, "Progress Report on Water Resources Program," April 14, 1954, 3, RG 115, Technical Correspondence, Agreements, & Reports (1944–1990), box 45, FRC, NARA (Denver).

27. Clark, "Summary of Activities," 1.

28. Dallas Watkins, "Monthly Summary of Activities," January 21, 1954, RG 115, Technical Correspondence, Agreements, & Reports (1944–1990), box 45, FRC, NARA (Denver).

29. Thomas A. Clark, letter to Marcus Gordon (Country Director, USO Ethiopia) re: "Tentative Program for a Survey of the Water Resources of the Awash River Basin," March 27, 1953, RG 115, International Reports (1914–1987), box 4, NARA (Denver).

30. US Bureau of Reclamation, "Water Resources—Ethiopia—Project 2, Monthly Summary of Activities," April 10, 1953, RG 115, box 6 (Communications, 1930–1960), FRC, NARA (Denver).

31. Thomas A. Clark, letter to H. E. Ato Getahoun Tessema, Addis Ababa, July 27, 1954, RG 59, Central Decimal File (1950–1954), box 993, National Archives and Research Administration at College Park, MD [hereafter NARA (College Park)].

32. Marcus J. Gordon, Memorandum to US State Department, "Water Resource Development Programs—Ethiopia," Washington, DC, May 28, 1954, RG 59, Central Decimal File (1950–1954), box 993, NARA (College Park).

33. Clark, letter to Marcus Gordon, 2–3.

34. The Ethiopian state also envisioned the Awash project as an exercise in comprehensive river basin development. The government commissioned a survey of the basin's development potential in 1961, which was funded and carried out by the United Nations Special Fund, one of the UN's technical assistance organizations. That survey, completed in 1965, involved a comprehensive series of studies detailing the water

resources, storage possibilities, soil and land classification, agronomic situation, dam sites, and geology of the entire watershed. The UN team envisioned over a dozen water storage dams and other diversion schemes that would have completely controlled the flows of the Awash River and, in theory, greatly enhanced the territory available for irrigated agriculture at a total cost of roughly US$14.5 million. The financing of the Awash II and Awash III dams, located just upstream of the Koka, was provided by the World Bank at a total cost of US$34.9 million. See United Nations Special Fund, *Report on Survey of the Awash River Basin, General Report* (Rome: FAO, 1965), 15–19.

35. "Address of Emperor Haile Selassie When Inaugurating Koka Dam," *Ethiopia Observer* 4 (1960): 247–48.

36. Theodore M. Vestal, *The Lion of Judah in the New World: Emperor Haile Selassie of Ethiopia and the Shaping of Americans' Attitudes toward Africa* (Santa Barbara, CA: Praeger, 2011), 105–11.

37. US Bureau of Reclamation, *Ethiopia–United States Cooperative Program for the Study of Water Resources—1961 Annual Report* (Washington, DC: US Bureau of Reclamation, October 31, 1961), 1.

38. Declan Conway, "A Water Balance Model of the Upper Blue Nile in Ethiopia," *Hydrological Sciences Journal* 42 (1997): 265–86.

39. See Conway, "Water Balance Model"; and Seleshi Awulachew et al., *A Review of Hydrology Sediment and Water Resource Use in the Blue Nile Basin* (Colombo, Sri Lanka: International Water Management Institute, 2008).

40. US Bureau of Reclamation, *Land and Water Resources of the Blue Nile Basin, Ethiopia* (Washington, DC: US Bureau of Reclamation, 1964), 4.

41. Ibid.

42. Cited in Alfred R. Golze, "Multipurpose Investigation of the Blue Nile," *Civil Engineering* 696 (1959): 40.

43. Golze, "Multipurpose Investigation," 41. The training of engineers and creation of a centralized water agency was characteristic of the Litani case as well as other Bureau overseas programs (including, prominently, Mexico, Turkey, Thailand, and Iran).

44. Anyone familiar with the history of water resource development in the United States cannot help note the irony in this perspective. While the United States does not have a public works bureaucracy, both the Bureau and the Army Corps of Engineers served as de facto agencies of public works throughout the twentieth century, and their collective waterway development projects—designed, studied, and paid for by the federal government—represent a staggering amount of public investment, often to the benefit of private enterprise. Still, the debate over public versus private roles in water resource development in the context of the Bureau's overseas interventions parallels similar domestic debates within the United States during the twentieth century. See Reisner, *Cadillac Desert*; Donald J. Pisani, *Water and American Government: The Reclamation Bureau, National Water Policy, and the West, 1902–1935* (Berkeley: University of California Press, 2002); and Billington and Jackson, *Big Dams*.

45. See John H. Spencer, *Ethiopia, the Horn of Africa, and U.S. Policy* (Cambridge, MA: Institute for Foreign Policy Analysis, 1977), 22–26; and Westad, *Global Cold War*, 253–61.

46. Operations Coordinating Board, "NSC Progress Report on U.S. Policy towards Ethiopia," December 18, 1957; Declassified Documents Reference System (hereafter DDRS), http://galenet.galegroup.com/servlet/DDRS (Doc. No. CK3100208768).

47. Edwin T. Layton, Memorandum for Chairman, Joint Chiefs of Staff, Washington, DC, June 27, 1956, DDRS, http://galenet.galegroup.com/servlet/DDRS (Doc. No. CK3100424574).

48. James H. Rives (African Operations), letter to Marcus J. Gordon, December 16, 1952, 4, RG 59, Central Decimal File (1950–1954), box 993, NARA (College Park). Rives made these calculations after a "quickie" review of the Bureau's preliminary reports on the Blue Nile, and he argued to State that the "impact of a full development [of the Blue Nile] upon the North African economy would be very great" due to Ethiopia's prodigious irrigation and hydropower potential.

49. Edward R. Felder (Chief, Ethiopia Branch, Technical Cooperation Administration), letter to Marcus Gordon, December 22, 1952, RG 59, Central Decimal File (1950–1954), box 993, NARA (College Park).

50. Layton, Memorandum for Chairman.

51. See Gabriel Kolko, *Another Century of War?* (New York: New Press, 2002), 26–27.

52. William Elliott, Memorandum for C. Herter, Office of the Special Assistant for National Security Affairs, "Preliminary Study of the Nile Watershed Necessary to the Determination of Some Important Aspects of United States Policy," February 14, 1958, DDRS, http://galenet.galegroup.com/servlet/DDRS (Doc. No. CK3100180009).

53. Christian Herter, letter to James S. Smith, Jr., Director of the International Cooperation Administration, February 21, 1958, DDRS, http://galenet.galegroup.com/servlet/DDRS (Doc. No. CK3100180009).

54. As noted in chapter 6, plans are currently under way within Ethiopia to investigate construction of several of the larger dams on the main channel of the Abay (Blue Nile) River and some more modest schemes on certain tributaries. Nearly all the proposed projects are multipurpose—with both irrigation and hydroelectric components—and are derived directly from the Bureau's 1964 report. See Paul J. Block, Kenneth Strzepek, and Bajali Rajagopalan, *Integrated Management of the Blue Nile Basin in Ethiopia: Hydropower and Irrigation Modeling* (Washington, DC: International Food Policy and Research Institute, May 2007).

55. US Bureau of Reclamation, *Land and Water Resources of the Blue Nile Basin.*

56. Tom A. Clark, Memorandum to Marcus Gordon, "Confidential Information Regarding Nile Basin Reconnaissance," June 5, 1952, RG 59, Central Decimal File (1950–1954), box 993, NARA (College Park).

57. In the early decades of its existence, the Bureau assigned project managers whose tasks, in addition to the obvious technical work needed to complete and operate dams and irrigation works, included "encouraging families of diverse backgrounds and farming experience to abide by existing regulations," resolving "petty squabbles among irrigators," and presiding over "project conferences that enabled settlers to air complaints." See Robinson, *Water for the West*, 33–35.

58. See Josephson, "'Projects of the Century.'"

59. Layton, Memorandum for Chairman.

60. L. W. Damours, Memorandum to Commissioner (Dominy) re: Foreign Activities—Annual Inspection—Blue Nile Project, Ethiopia, Project #3, November 13, 1959, Floyd E. Dominy papers (1934–2002), box 22, AHC (Laramie).

61. Floyd Dominy, letter to the Secretary of the Interior re: "Report of Inspection, Blue Nile Project, Ethiopia," December 30, 1959, Floyd E. Dominy papers (1934–2002), box 22, American Heritage Center (hereafter AHC), University of Wyoming, Laramie.

62. Donald Barnes (Project Engineer), "Progress in Project Relationships, USOM/E—IEG—Foreign Activities—Ethiopia, Project No. 3," September 5, 1960, RG 115, Technical Correspondence, Agreements, & Reports (1944–1990), box 45, NARA (Denver).

63. Damours, Memorandum to Commissioner (Dominy).

64. R. Fisher, "Revision of Blue Nile Project to Include Mapping," August 21, 1959, 2, RG 469.6, Records of the International Cooperation Administration (hereafter ICA) (1951–1961), General Correspondence, box 7, NARA (College Park).

65. Ibid., 8.

66. Alfred R. Golze, Confidential Airmail to Donald Barnes, Project Engineer (Blue Nile Investigations), September 13, 1960, 2, RG 115, Technical Correspondence, Agreements, & Reports (1944–1990), box 45, FRC, NARA (Denver). This information was delivered in a "Blue Envelope" (i.e., highly confidential) communication internal to Bureau staff, who were formulating a response to chastisement by embassy personnel for their attitudes toward technical assistance in Ethiopia.

67. Ibid.

68. Ministry of Foreign Affairs (IEG), letter to United States Embassy, May 23, 1952, RG 469.2, Records of the Economic Cooperation Administration, General Correspondence (1945–1952), box 14, NARA (College Park).

69. The presence of such attitudes is evidenced by a photograph of Floyd Dominy's visit accompanying the Blue Nile report. A handwritten list of the people featured on the back of the image listed no Ethiopian personnel by name, instead referring to them as "Ethiopian official" or similarly nondescript labels.

70. Donald Barnes, Blue Envelope letter to D. W. Damours, October 19, 1960, RG 115, Technical Correspondence, Agreements, & Reports (1944–1990), box 45, FRC, NARA (Denver).

71. Timothy Mitchell makes a similar point regarding the Aswan High Dam on the Nile River in Egypt, discussing how this massive scheme—in contrast to traditional water distribution systems in the basin that were organized around hundreds of small weirs, ditches, dikes, and canals—concentrates the observer's gaze and intensifies the sense that technology has conquered nature. See Mitchell, *Rule of Experts*, 35–36.

72. Homer Bigart, "U.S. Is Regaining Ethiopia's Favor," *New York Times*, February 18, 1957, 3–4.

73. Confidential Memorandum (IGATO—A-345), January 16, 1959, RG 59, Central Decimal File (1955–1959), box 722, NARA (College Park).

74. Ibid.

75. USOM, Airgram to ICA/Washington, "Blue Nile Survey Project," March 9, 1959, RG 59, Central Decimal File (1950–1954), box 722, NARA (College Park).

76. Ralph H. Fisher, Program Officer, "Blue Nile Survey," January 5, 1959, RG 59, Central Decimal File (1950–1954), box 722, NARA (College Park)

77. Drew Pearson, "Ethiopia Still Waiting for U.S. Dam," *Washington Post Times Herald*, June 8, 1964, B27.

78. Ibid.

79. American Embassy [Addis Ababa], "Drew Pearson's Articles on Attempt to Elicit More US Material Aid for Ethiopia," July 4, 1964, DDRS, http://galenet.galegroup .com/servlet/DDRS (Doc. No. CK3100357404).

80. C. L. Schultze, Memorandum to the President, June 4, 1966, RG 59, Subject-Numeric Files (1964–1966), box 710, NARA (College Park).

81. David S. Bell, "New Project Approval—Finchaa Hydroelectric Project in Ethiopia," Memorandum for the President, May 17, 1966, 1–2, DDRS, http://galenet.galegroup .com/servlet/DDRS (Doc. No. CK3100423822).

82. Ibid., 4.

83. See N. Kotler, "The Over-Present Americans," *Nation*, February 20, 1967, 236–39. The communications facility was part of a military aid package negotiated by then–Vice President Richard Nixon during an official visit in 1957.

84. See Kolko, *Confronting the Third World*; and Westad, *Global Cold War*.

CHAPTER FIVE

1. The anecdote and quotes are from Floyd Dominy, transcript of Bureau of Reclamation oral history interviews conducted and edited by Brit Allan Storey, April 6, 1994, and April 8, 1996, NARA (College Park), 76–77.

2. My argument regarding Pa Mong's influence irrespective of its existence resonates strongly with Jonathan Peyton's deployment of an intriguing notion, the "unbuilt environment," first used by historian Kathryn Oberdeck. Peyton traces the conception and undertaking of a series of data collection activities by BC Hydro, a prominent Canadian hydropower firm, in the wake of a failed effort to construct five large dams on the Stikine and Iskut Rivers. Although dam-induced river basin transformation never actually occurred, Peyton argues convincingly that the environments of the basins nevertheless changed—materially and discursively—as a result of the wide variety of actors mobilized both in support of and in opposition to the planned developments. See Jonathan Peyton, "Corporate Ecology: BC Hydro's Stikine-Iskut Project and the Unbuilt Environment," *Journal of Historical Geography* 37 (2011): 369. Thanks are due an anonymous reviewer who directed me toward this work.

3. Dominy, oral history interviews, 77–78. The personal and political foibles of Floyd Dominy earned him an entire chapter in Marc Reisner's magnum opus. See Reisner, *Cadillac Desert*, 214–54. The commissioner eventually became the nemesis of American environmentalist David Brower in early debates over the environmental impacts of large dams in the western United States. See John McPhee, *Encounters with the Archdruid* (New York: Farrar, Straus and Giroux, 1977).

4. For an enlightening history of the role of Mekong development initiatives in the regional geopolitics of the Cold War, see Thi Dieu Nguyen, *The Mekong River and the Struggle for Indochina: Water, War, and Peace* (Westport, CT: Praeger, 1999). David Biggs highlights how the Bureau became engaged in nation-building efforts in Southeast Asia due to the "unique circumstances of the Cold War in Indochina" and offers a detailed accounting of the Bureau's activities in light of encounters with their counterparts in the Thai and Lao governments. In addition, he astutely points out that the Bureau and its technical knowledge did not displace, but rather extended, the reach of "more local perspectives" on modernization in the region, and that it ultimately transferred the "American engineers' strategy of claiming political immunity" by falling back on the supposed objectivity of technical knowledge. See Biggs, "Reclamation Nations," 227–42. In elucidating the manner in which the Bureau advanced ideas of modernization in the region, however, he pays less attention to the roles of technology and technological expertise, and their entanglements in America's geopolitical vision, in the efforts to transform water governance in the Mekong.

5. See Bruno Latour, "Drawing Things Together," in *Representation in Scientific Practice*, ed. Michael Lynch and Steven Woolgar (Cambridge, MA: MIT Press, 1990), 40–45.

6. There is a now considerable literature on the hydropolitics of the Mekong River basin, and I return to the contemporary geopolitics of dam construction in the Mekong in the following chapter. For representative examples, see Karen Bakker, "The Politics of Hydropower: Developing the Mekong," *Political Geography* 18 (1999): 209–32; Chris Sneddon and Coleen Fox, "Rethinking Transboundary Waters: A Critical Hydropoli-

tics of the Mekong Basin," *Political Geography* 25 (2006): 181–202; Philip Hirsch, "Water Governance Reform and Catchment Management in the Mekong Region," *Journal of Environment & Development* 15 (2006): 184–201; and R. Edward Grumbine, John Dore, and Jianchu Xu, "Mekong Hydropower: Drivers of Change and Governance Challenges," *Frontiers in Ecology and the Environment* 10 (2012): 91–98.

7. US Bureau of Reclamation, *Reconnaissance Report: Lower Mekong River Basin*, prepared for International Cooperation Administration (United States Bureau of Reclamation, March 1956).

8. Past and recent academic work on Mekong development has often failed to appreciate the extent of US involvement in establishing the institutional parameters for cooperative water resource development in the mid-1950s. A great deal of work continues to emphasize the role of the United Nations' Economic Committee for Asia and the Far East (ECAFE), and of the UN more generally, in prompting creation of the Mekong Agreement of 1957. This omission seriously underplays US strategic objectives in fostering directly, via its embassy personnel and the Bureau survey, regional cooperation in mainland Southeast Asia and likewise neglects the ways in which UN initiatives were influenced by and reflective of US geopolitical priorities during this era.

9. Officer in Charge of Thai and Malayan Affairs (Landon), Memorandum of Conversation, July 20, 1954, 647, in *FRUS, 1952–1954*, vol. 12, part 1, *East Asia and the Pacific* (Washington, DC: GPO, 1954), 645–47.

10. J. Mecklin, Dispatch from Saigon, January 27, 1955, RG 59, Central Decimal Files (1955–1959), box 29 (Mekong Survey), NARA (College Park).

11. H. E. Stassen, Telegram Sent to Bangkok, Vientiane, Phnom Penh, Saigon, "Subject: Mekong Survey," May 13, 1955, RG59, Central Decimal Files (1955–1959), box 29 (Mekong Survey), NARA (College Park).

12. Gilbert M. Strauss, Joint Embassy/United States Operations Mission (USOM) Telegram to Secretary of State, June 17, 1955, RG 59, Central Decimal Files (1955–1959), box 29 (Mekong Survey), NARA (College Park).

13. The chief of the team, Robert Newell, was a former regional director of the Bureau in the Pacific Northwest; H. Velpeau Darling and George Tomlinson were widely respected engineers and had worked in partnership with the Corps of Engineers and the Tennessee Valley Authority, respectively.

14. United Nations Economic Commission for Asia and the Far East, *Development of Water Resources in the Lower Mekong Basin* (Bangkok: ECAFE, 1957).

15. Memorandum to L. Metcalfe Walling, Director of the United States Economic Aid Mission to Cambodia, January 27, 1955, RG 59, Central Decimal Files (1955–1959), box 29 (Mekong Survey), NARA (College Park).

16. US Bureau of Reclamation, *Lower Mekong River Basin*, i.

17. Ibid.

18. Telegram from USOM (Bangkok) to ICA (Washington, DC), February 1, 1956, RG 469.7, Records of the ICA (1951–1961), box 12, NARA (College Park).

19. National Security Council (NSC), "U.S. Policy in the Far East," September 25, 1959, DDRS, http://galenet.galegroup.com/servlet/DDRS (Doc. No. CK3100167891).

20. Mekong Committee, *Brief Description of the Pa Mong Project* (Bangkok: Economic Commission for Asia and the Far East, June 30, 1961).

21. Schaaf and Dominy cultivated a personal relationship as well. A letter from Schaaf to Dominy regarding the Bureau's work in the Mekong region in the late 1960s included a handwritten note regarding a potential golf outing during Schaaf's next visit to the United States. See Floyd E. Dominy papers (1934–2002), box 29, AHC (Laramie).

22. B. F. Dixon, Memorandum to State Department, "Visit of the U.S. Commissioner of Reclamation Floyd E. Dominy to Inspect the Pa Mong area of the Mekong Development Scheme," July 14, 1961, RG 59, Central Decimal Files (1960–1963), box 14, NARA (College Park). Dominy was initially hesitant to commit the Bureau to the Mekong studies, as he believed that a private engineering firm might be better qualified to undertake the investigation. As Biggs discusses, the Bureau's policy with regard to overseas activities was to focus solely on feasibility and other preliminary studies, and on occasion project design, and to encourage the use of private firms (hopefully American, according to State Department discussions) for actual construction. See Biggs, "Reclamation Nations," and also chapters 3 and 4 of the present work.

23. Dixon, Memorandum to State Department. The great irony here is that Dominy's resume included almost no formal training as an engineer, and he was at best condescending toward the engineers and other water experts under his direction at the Bureau throughout the 1960s. Dominy's real genius, as Reisner's work makes so abundantly clear, was in bureaucratic politics and promoting the Bureau's organizational interests. See Reisner, *Cadillac Desert*, 214–54. Robinson describes him as "the most dynamic, two-fisted, and controversial individual to head the agency." See Robinson, *Water for the West*, 82.

24. Kenneth P. Landon, Memorandum to Walt W. Rostow, "Political Implications of the Mekong River Basin for Southeast Asia," March 6, 1961, DDRS, http://galenet .galegroup.com/servlet/DDRS (Doc. No. CK3100262618). In terms of prioritizing political justifications for a large dam in the face of technical and environmental uncertainties, the parallels to current debates over two major hydroelectric projects being contemplated on the Mekong River within the Lao PDR, wherein a variety of scientists and advocacy organizations have criticized the Lao government's poorly conceived environmental impact assessments, are striking (see chap. 6).

25. US Bureau of Reclamation, *Mekong Pa Mong Survey (Phase I): Interim Report* (Denver: US Bureau of Reclamation, June 1965).

26. US Bureau of Reclamation, *Pa Mong Project (Lower Mekong River Basin), Phase I Report*, vol. 2, appendix I (Agreements) (Denver: United States Bureau of Reclamation, March 1966), 4.

27. US Bureau of Reclamation, *Pa Mong Project, Stage One, Phase II (Executive Summary)* (Denver: United States Bureau of Reclamation, 1970), 8.

28. Mekong Committee, *Brief Description*.

29. US Bureau of Reclamation, *Mekong Pa Mong Survey (Phase I)*, 20.

30. United States Agency for International Development, *To Tame a River* (Washington, DC: USAID, 1968), 13.

31. Karabell, *Architects of Intervention*, 206–24.

32. Dixon, Memorandum to State Department.

33. Lyndon Baines Johnson, "Johnson on Southeast Asian Aid," *Current History* 49 (1965): 303.

34. Lyndon Baines Johnson, "Peace without Conquest" (address at Johns Hopkins University, April 7, 1965), in *Public Papers of the Presidents of the United States, Lyndon B. Johnson, Containing the Public Messages, Speeches, and Statements of the President, 1965, Book I—January 1 to May 31, 1965* (Washington, DC: US Government Printing Office, 1966), 397.

35. Willard Hanna, *The Prize at Pa Mong*, part 5 of *The Mekong Project, AUFS Southeast Asia Series*, vol. 16, no. 14 (Hanover, NH: American Universities Field Staff, 1968), 1.

36. Hanna, *Prize at Pa Mong*, 5.

37. Latour, "Drawing Things Together," 67.

38. According to Willard Hanna, a long-term observer of American development assistance in the region, Pa Mong was "no $6 to $7 billion, ten-year project to which the United States can generously contribute a billion or perhaps two billion and then gracefully withdraw. This is at least a $50 billion, twenty-five year project, in which relative success will mean continuing escalation of the American investment, and relative failure could mean total loss, to Americans at least, of any advantage." See Hanna, *Prize at Pa Mong*, 11.

39. As early as 1962, a preliminary study of the Pa Mong catchment area identified the suitability of land for irrigation in the project area as a "critical problem," noting that Pa Mong's "irrigation aspects . . . do offer definite limitations in the scope of the Pa Mong Project as shown by the general paucity of good irrigable areas . . . particularly in Northeast Thailand." See US Bureau of Reclamation, *Preliminary Study, Pa Mong Project, Thailand-Laos* (Denver: United States Bureau of Reclamation, January 1962), 29–30. Another Bureau expert noted that the project area's soils were "beset with serious problems involving extremely low fertility and highly permeable top soil underlain by impermeable substrata." See US Bureau of Reclamation, *Preliminary Study, Pa Mong Project*, appendix B, 2.

40. Dean Rusk, Telegram re: "Pa Mong Investigations," March 19, 1968, RG 59, Subject-Numeric Files (1967–1969), box 642, NARA (College Park).

41. Rutherford Poats, Memorandum to Jack Valenti, "Lower Mekong Basin Development Scheme," March 26, 1965, DDRS, http://galenet.galegroup.com/servlet/DDRS (Doc. No. CK3100409056).

42. See John Lewis Gaddis, *Strategies of Containment: A Critical Appraisal of American National Security Policy during the Cold War*, 2nd ed. (New York: Oxford University Press, 2005), 272–306.

43. W. Woodbury, "Planning for Post War Reconstruction and Development in Southeast Asia," May 31, 1972, RG 59, Subject-Numeric Files (1970–1973), box 714, NARA (College Park).

44. "Mekong Activities—aide-memoire," August 20, 1970 (attached to State Department Memorandum re: World Bank Proposal to Support Mekong Development, September 17, 1970), RG 59, Subject-Numeric Files (1970–1973), box 714, NARA (College Park).

45. Ibid.

46. Ibid.

47. Leonard Unger, Airgram to Department of State, "Mekong Committee Statements at ECAFE 26th Plenary Session, Bangkok, April 24, 1970," May 7, 1970, RG 59, Subject-Numeric Files (1970–1973), box 714, NARA (College Park).

48. See the seminal papers delivered at a conference in December 1968, collected in M. Taghi Farvar and John P. Milton, eds., *The Careless Technology: Ecology and International Development* (Garden City, NY: Natural History Press, 1972).

49. See Warne, *Bureau of Reclamation*, 207; and Robinson, *Water for the West*, 92–95.

50. Claire Sterling, "Thai-Laos Dam Plan Is Perfect One—Except for Why?," *Washington Post*, May 1, 1971, A18.

51. In an earlier and remarkably prescient article, Sterling detailed the potential ecological impacts of Mekong basin development schemes: the loss of silt in the delta region due to upstream dams and the resultant effects on soil fertility in this critical rice-growing area; the blockage of nutrients and resultant impacts on economically important fish populations and other aquatic organisms; the extinction of riverine

organisms, including the Mekong giant catfish; damage to the Tonle Sap fisheries, on which millions of people depended for food and income; the efflorescence of invasive weeds in newly created reservoirs; salinization in those areas of Northeast Thailand that are underlain by an enormous salt formation by irrigation with Mekong waters; the inundation of high-quality agricultural lands; and resettlement of millions of inhabitants and increases in waterborne diseases due to reservoir creation. See Claire Sterling, "40-Odd Dams Hold Promise for Great Mekong Basin," *Washington Post*, April 24, 1971, A18.

52. A number of State Department memoranda and telegrams from roughly 1970 to 1974 drive home this point.

53. Woodbury, "Planning for Post War Reconstruction."

54. See Chris Sneddon and Coleen Fox, "Water, Geopolitics, and Economic Development in the Conceptualization of a Region," *Eurasian Geography and Economics* 53 (2012): 143–60.

55. See Chris Sneddon and Coleen Fox, "Inland Capture Fisheries and Large River Systems: A Political Economy of Mekong Fisheries," *Journal of Agrarian Change* 12 (2012): 279–99.

CHAPTER SIX

1. See Serres, *Conversations*, 60.

2. Occasionally new rationales are provided for old dams as well. During the course of my fieldwork in the late 1990s on the Nam Phong Dam in Northeast Thailand—constructed in the mid-1960s under the auspices of the Mekong Committee as a "pilot" project on a prominent tributary system of the Mekong River—I was surprised to hear several Thai officials from prominent water resource agencies claim that the dam's primary functions were storage for irrigation and flood control; one official claimed that electricity was a "by-product" of the dam's implementation. My research on the history of the project clearly showed that the government's primary goal in building the Nam Phong scheme was the production of hydroelectricity to meet the nation's burgeoning demand in the 1960s. Since the amount of electricity produced by the project's power station (roughly 25 megawatts) has been minuscule in terms of total demand in the three decades since its construction, government officials simply reinterpreted the dam's purpose to fit the development narrative they desired. This example shows the highly adaptive nature of dams as development agents—whose "actions" and effects can ignore original intentions—and underscores the need for genealogical research.

3. For a different yet complementary angle on the current geopolitics of large dams, see Desiree Tullos, Bryan Tilt, and Catherine Reidy Liermann, "Introduction to the Special Issue: Understanding and Linking the Biophysical, Socioeconomic and Geopolitical Effects of Dams," *Journal of Environmental Management* 90 (2009): S203–S207.

4. Within the litany of large-dam projects that were ill conceived from inception and have perpetrated a host of socioecological disruptions, the Cahora Bassa stands out as one of the most egregious. For a superb history of this project, see Isaacman and Isaacman, *Dams, Displacement, and the Delusion of Development*.

5. The complexities of the political, economic, and ideological dynamics, interacting across a six-decade time span, that gave rise to the Three Gorges Dam on nearly the precise spot surveyed by John Savage in 1943 (see chap. 2) are surely worth pursuing in alignment with this book's other primary cases, but are beyond its scope and are covered in other work. See Fearnside, "China's Three Gorges Dam"; and Boxer,

"China's Three Gorges Dam." For a popular and well-researched account, see Chet-ham, *Before the Deluge.*

6. See Michael Hammond, "The Grand Ethiopian Renaissance Dam and the Blue Nile: Implications for Transboundary Water Governance," GWF Discussion Paper 1307 (Canberra: Global Water Forum, February 2, 2013), http://www.globalwaterforum.org/2013/02/18/the-grand-ethiopian-renaissance-dam-and-the-blue-nile-implications-for-transboundary-water-governance/.

7. Economist Intelligence Unit, "Ethiopia Alternatives: Dam Bond Launched," October 24, 2011, http://www.energyrealities.org/content/ethiopia-alternatives-dam-bond-launched/erp3A257294F5E318FE8.

8. Associated Press, "In Ethiopia, a Massive Nile River Dam Project Inspires Comparison with the Story of Hoover Dam," July 2, 2013, http://www.foxnews.com/world/2013/07/02/in-ethiopia-massive-nile-river-dam-project-inspires-comparison-with-story/.

9. Ashok Swain, "Ethiopia's Dam Project Reflects Shifting Balance of Power in Nile Basin," *WPR Briefing* 982 (June 13, 2013), http://www.worldpoliticsreview.com/authors/982/ashok-swain.

10. Ashok Swain, "Challenges for Water Sharing in the Nile Basin: Changing Geo-Politics and Changing Climate," *Hydrological Sciences Journal* 56 (2011): 700.

11. See Daniel Berhane, "Ethiopia: Fincha Hydroelectric Plant Becomes Operational," *Horn Affairs*, December 11, 2011 (6:02 a.m.), http://danielberhane.com/2011/12/20/ethiopia-fincha-hydropower-plant-becomes-operational/.

12. See Michael Rubin, "Iran's Dam Diplomacy," *Operation Environment Watch*, American Enterprise Institute, May 7, 2013, http://www.aei.org/article/foreign-and-defense-policy/regional/middle-east-and-north-africa/irans-dam-diplomacy/; and Neil Mac-Farquhar, "Iran Is Seeking Lebanon Stake as Syria Totters," *New York Times*, May 24, 2012, http://www.nytimes.com/2012/05/25/world/middleeast/with-syria-in-turmoil-iran-seeks-deeper-partner-in-lebanon.html?pagewanted=all&_r=0. The central irony of the Iranian government's forays into "dam diplomacy" is the occurrence within their own borders of a "water super-crisis" so severe that it threatens basic food production systems, a situation likely to be exacerbated by population growth and climate change. See Masoud Yazdanpanah et al., "A New Enemy at the Gate: Tackling Iran's Water Super-Crisis by Way of Transition from Government to Governance," *Progress in Development Studies* 13 (2013): 178.

13. See US Bureau of Reclamation, *Water Resources Investigations for the Nahr El-Barad Basin, Reconnaissance Report* (Denver: United States Bureau of Reclamation, September 1956).

14. Stephen Dockery, "Lack of Dams Holding Back Lebanon's Water, Energy Sectors," *Daily Star* [Lebanon], March 2, 2013, http://www.dailystar.com.lb/News/Local-News/2013/Mar-02/208501-lack-of-dams-holding-back-lebanons-water-energy-sectors .ashx#axzz2gNzdsAEO. Although claims of water shortage are not to be dismissed, it is equally important to recognize the ways in which social actors, particularly governments, construct scarcity through discourses and policies that always mediate instances of physical scarcity. See Alatout, "'States' of Scarcity," 961; and Basil Mahayni, "Producing Crisis: Hegemonic Debates, Mediations and Representations of Water Scarcity," in *Contemporary Water Governance in the Global South: Scarcity, Marketization and Participation*, ed. Leila Harris, Jacqui Goldin, and Christopher Sneddon (New York: Routledge, 2013), 35–44.

15. See Mark Zeitoun et al., "Hydro-Hegemony in the Upper Jordan Waterscape: Control and Use of the Flows," *Water Alternatives* 6 (2013): 86–106.

16. See Philip Hirsch, "China and the Cascading Geopolitics of Lower Mekong Dams," *Asia-Pacific Journal* 9 (2011): 1–4.

17. See J. Ferrie, "Laos Turns to Hydropower to Be 'Asia's Battery,'" *Christian Science Monitor*, July 2, 2010, http://www.csmonitor.com/World/Asia-Pacific/2010/0702/Laos-turns -to-hydropower-to-be-Asia-s-battery.

18. See Edward Grumbine and Jianchu Xu, "Mekong Hydropower Development," *Science* 332 (2011): 178–79.

19. See Philip Hirsch, "The Changing Political Dynamics of Dam Building on the Mekong," *Water Alternatives* 3 (2010): 318–19.

20. The headline of a recent article in a Japanese daily reads, "China Using Development Aid to Spearhead Asia Push." See Y. Nishii and F. Go, *Asahi Shimbun*, January 26, 2011, http://www.cam111.com/photonews/2011/01/26/74545.html.

21. Quoted in "Lower Mekong Initiative," *Voice of America*, July 31, 2010, http://editorials .voa.gov/content/lower-mekong-initiative-99780124/1481965.html.

22. Simon Roughneen, "US Dips into Mekong Politics," *Asia Times Online*, August 14, 2010, http://www.atimes.com/atimes/Southeast_Asia/LH14Ae01.html. Security analyst Richard Cronin argues that "all of the Lower Mekong countries understand the geopolitical nature of the U.S. initiative, most especially China, and to varying degrees and the exception of China, they all welcome it" due to concerns over "China's hegemonic potential both in mainland Southeast Asia and the South China Sea." See Richard P. Cronin, statement before the Senate Committee on Foreign Relations, Hearing on Challenges to Water and Security in Southeast Asia, September 23, 2010, 7, http://www.stimson.org/summaries/testimony-before-senate-foreign-relations-hearing -on-water-and-security-in-southeast-asia/.

23. Cronin, statement, 9–10.

24. "Four Asian Countries Including Pakistan Will Build Over 400 Dams," *News*, August 12, 2013, http://www.thenews.com.pk/article-113272-Four-Asian-countries -including-Pakistan-will-build-over-400-dams.

25. R. Edward Grumbine and Maharaj K. Pandit, "Threats from India's Himalaya Dams," *Science* 339 (January 4, 2013), 36.

26. Rachel Beitarie, "Surge of Dams in Southwest China Produces Power and Public Ire," *Circle of Blue*, November 30, 2010, http://www.circleofblue.org/waternews/2011/ world/burst-of-new-dams-in-southwest-china-produces-power-and-public-ire/.

27. Matt Finer and Clinton N. Jenkins, "Proliferation of Hydroelectric Dams in the Andean Amazon and Implications for Andes-Amazon Connectivity," *PLoS ONE* 7 (2012): 2, doi: 10.1371/jounal.pone.35126.

28. See Michael Goldman, *Imperial Nature: The World Bank and Struggles for Justice in the Age of Globalization* (New Haven, CT: Yale University Press, 2005).

29. John Briscoe, "The Changing Face of Water Infrastructure Financing in Developing Countries," *International Journal of Water Resources Development* 15 (1999): 301–8.

30. World Bank Group, *Directions in Hydropower*, 4.

31. Ibid., 6.

32. Ibid., 12.

33. Quoted in Howard Schneider, "World Bank Turns to Hydropower to Square Development with Climate Change," *Washington Post*, May 8, 2013, http://www.washingtonpost .com/business/economy/world-bank-turns-to-hydropower-to-square-development -with-climate-change/2013/05/08/b9d60332-b1bd-11e2-9a98-4be1688d7d84_ print.html.

34. Emmanuel Grenier, "The WCD Question in Marseille (March 16, 2012)," *News*, International Commission on Large Dams, http://www.icold-cigb.org/gb/news/news.asp?IDA=254.

35. The World Bank has proved particularly deft at combining discourses of "water for all," participatory water management, poverty alleviation, and the continuing need for large-scale water development within its loan portfolios. See Goldman, *Imperial Nature*.

36. For a more detailed accounting of the rise of hegemonic concepts within institutions of water governance, particularly as applied in the Global South, see Chris Sneddon, "Water, Governance and Hegemony," in *Contemporary Water Governance*, ed. Harris, Goldin, and Sneddon, 13–24.

37. The WCD also followed closely on the heels of declarations by water resource agencies in the United States that large hydroelectric and water storage dams were no longer part of the discussion of how to best use the country's water resources. No less a personage than Commissioner of Reclamation Daniel Beard stated that large dams "are too expensive" and not worth the environmental and social costs. The "era of big dams," declared Beard, "is over." The tremendous irony of Beard's statements, neglected by the numerous commentaries that quote his declaration, is that the commissioner was responding to a question regarding the Three Gorges Dam, about which he stated, "There is no more visible symbol in the world of what we [the United States] are trying to move away from." One wonders how John Savage might have responded upon hearing this news. Beard is quoted in Simon Winchester, *River at the Center of the World: A Journey Up the Yangtze, and Back in Chinese Time* (New York: Picador, 2004), 228.

38. There is a significant risk of overstating and oversimplifying China's influence within the global capitalist economic system, a risk exacerbated by hyperbolic media reports. For an astute assessment of the links between "China's rise" and the most recent iteration of economic globalization, see Padraig Carmody, Godfrey Hampwaye, and Enock Sakala, "Globalisation and the Role of the State? Chinese Geogovernance in Zambia," *New Political Economy* 17 (2012): 209–10.

39. The scholarly work on this topic is interdisciplinary and wide ranging. For a general overview of the role of China and other Southern donors in altering global foreign aid dynamics, see Ngaire Woods, "Whose Aid? Whose Influence? China, Emerging Donors and the Silent Revolution in Development Assistance," *International Affairs* 84 (2008): 1205–21; Clemens Six, "The Rise of Postcolonial States as Donors: A Challenge to the Development Paradigm?," *Third World Quarterly* 30 (2009): 1103–21; and Patrick Kilby, "The Changing Development Landscape in the First Decade of the 21st Century and Its Implications for Developing Countries," *Third World Quarterly* 33 (2012): 1001–17. Moreover, there are sharp disagreements about what should be classified as "development" aid. The Development Assistance Committee (DAC), a small group of Western countries belonging to the OECD, defines aid quite narrowly under the phrase "official development assistance," while much of the aid from China and other emerging donors (e.g., export credits, non-concessional state loans, or funds used to bolster donor investment) is not considered "official" assistance. See Woods, "Whose Aid?," 1205–6; and Deborah Brautigan, "Aid with 'Chinese Characteristics': Chinese Foreign Aid and Development Finance Meet the OECD-DAC Regime," *Journal of International Development* 23 (2011): 752–64.

40. Emma Mawdsley identifies several common features of the economic assistance offered by Southern development actors, including a "developing country" identity

shared with recipient states, a refusal of donor-recipient hierarchies, an assertion of mutual benefits, and an appropriate level of technical expertise. She also assiduously debunks many of the problematic claims regarding more equitable and effective foreign aid that "new" donor nations profess. See Emma Mawdsley, "The Changing Geographies of Foreign Aid and Development Cooperation: Contributions from Gift Theory," *Transactions of the Institute of British Geographers* 37 (2012): 268–69.

41. For examples, see Marcus Power, Giles Mohan, and May Tan-Mullins, *China's Resource Diplomacy in Africa: Powering Development?* (New York: Palgrave Macmillan, 2012).

42. See Asia Sentinel, "China's Dam-Builders to the World," *Irrawaddy*, September 16, 2011, http://www2.irrawaddy.org/article.php?art_id=22083.

43. These numbers are taken from International Rivers, *The New Great Walls: A Guide to China's Overseas Dam Industry*, 2nd ed. (Berkeley, CA: International Rivers, November 2012), 4.

44. Ibid., 28–29. For a comprehensive summary of China's increasing interest in exporting hydropower development, see Kristen McDonald, Peter Bosshard, and Nicole Brewer, "Exporting Dams: China's Hydropower Industry Goes Global," *Journal of Environmental Management* 90 (2009): S294–S302.

45. See McDonald, Bosshard, and Brewer, "Exporting Dams," S297–S298. For an instructive case in Ghana, see Oliver Hensengerth, "Chinese Hydropower Companies and Environmental Norms in Countries of the Global South: The Involvement of Sinohydro in Ghana's Bui Dam," *Environment, Development and Sustainability* 15 (2013): 285–300.

46. There has been an efflorescence of articles and books in recent years on China's rising influence in the Global South and its novel approach to "aid diplomacy." In line with the cases presented in this book, Africa and Southeast Asia have been particularly prominent targets of Chinese economic and technical assistance (and investment) around hydropower development. For Africa, see Peter Kragelund, "Knocking on a Wide Open Door: Chinese Investments in Africa," *Review of African Political Economy* 36 (2009): 479–97; and Dominik Kopinski, Andrzej Polus, and Ian Taylor, "Contextualising Chinese Engagement in Africa," *Journal of Contemporary African Studies* 29 (2011): 129–36. For Southeast Asia, see Stephen Frost, "Chinese Outward Direct Investment in Southeast Asia: How Big Are the Flows and What Does It Mean for the Region?," *Pacific Review* 17 (2004): 323–40. For perhaps obvious reasons, China has a much longer history of trade and aid relations with Southeast Asia than do most Western donors.

47. This relationship, however, as in the case of the Bureau and the State Department, is not without tensions. The construction of major hydropower facilities within China is now highly competitive, often cutting across political and personal alliances. See John Dore, Yu Xiaogang, and K. Yuk-shing Li, "China's Energy Reforms and Hydropower Expansion in Yunnan," in *Democratizing Water Governance in the Mekong Region*, ed. Louis Lebel et al. (Chiang Mai, Thailand: Mekong Press, 2007), 55–92.

48. See Hirsch, "Cascading Geopolitics."

49. McDonald, Bosshard, and Brewer, "Exporting Dams," S300–S301.

50. The field of political ecology is now firmly established as a vital and constantly evolving framework for analyzing and interpreting human-environment relationships. For timely and comprehensive overviews of the history and current theoretical and practical engagements of the field, see Roderick P. Neumann, *Making Political Ecology* (London: Hodder Arnold, 2005); and Paul Robbins, *Political Ecology: A Critical Introduction*, 2nd ed. (Malden, MA: Wiley-Blackwell, 2012).

51. A good deal of the notoriety of large dams and their effects in the United States accrues to the work of David Brower and the Sierra Club, who launched an ultimately doomed campaign against the Bureau-designed Glen Canyon Dam on the Colorado River. See McPhee, *Encounters with the Archdruid*. It must be noted that Brower's criticisms of large dams rested almost entirely on a preservationist ethic, focusing on dams' environmental impacts. This focus stands in stark contrast with the work of International Rivers, affiliated rivers-based organizations throughout Asia, Africa, Europe, and Latin America, and the loose coalition of communities organized under the Dam Affected Peoples network, whose critique of large dams encompasses social justice and disrupted livelihoods in addition to ecological effects.

52. For an excellent analysis of this neglected component of the history of the dissemination of large dams, see Ann Danaiya Usher, "The Race for Power in Laos: The Nordic Connections," in *Environmental Change in South-East Asia: People, Politics and Sustainable Development*, ed. Michael J. G. Parnwell and Raymond Bryant (London: Routledge, 1996), 117–38.

53. For an empirically and theoretically rich account of the Pak Mun struggle, see Jim Glassman, "From Seattle (and Ubon) to Bangkok: The Scales of Resistance to Corporate Globalization," *Environment and Planning D: Society and Space* 19 (2001): 520–24. For the Narmada campaign, the seminal version is Khagram, *Dams and Development*, 65–138.

54. An excellent summary of this knowledge is found in McCully, *Silenced Rivers*, 29–100.

55. The WCD, launched in 1997, was a joint brainchild of the International Union for Conservation of Nature and the World Bank, and was sponsored by the United Nations Development Programme (UNDP). The actual report includes analysis of the economic, technical, environmental, and social aspects of the planet's roughly 45,000 large dams. In brief, the report concludes that although dams "have made an important and significant contribution to human development," they have too often produced severe social and environmental impacts borne disproportionately "by people displaced, by communities downstream, by taxpayers and by the natural environment." See WCD, *Dams and Development*, 310. Both the Sardar Sarovar and Pak Mun Dams were investigated by the WCD, and Medha Patkar, a leading spokesperson for the anti-Sardar Sarovar campaign in India, was a member of the commission.

56. Recently, a special issue of the journal *Water Alternatives* explored just this question, but came up with no conclusive answers. Not surprisingly, proponents of the WCD's critical evaluation of large dams see the report as an important way to legitimate the concerns and struggles of those disenfranchised by state-sponsored water resource development, while representatives of the dam industry and a wide range of government officials see the report as irrelevant for a world in great need of water infrastructure. See Moore, Dore, and Gyawali, "World Commission on Dams +10"; Navroz Dubash, "Reflections on the WCD as a Mechanism of Global Governance," *Water Alternatives* 3 (2010): 416–22; and Briscoe, "Overreach and Response."

57. For an overview of the global political dynamics of this issue and an excellent case applied to Costa Rica, see Robert Fletcher, "When Environmental Issues Collide: Climate Change and the Shifting Political Ecology of Hydroelectric Power," *Peace and Conflict Monitor* 5 (2011): 14–30.

58. See Vincent L. St. Louis et al., "Reservoir Surfaces as Sources of Greenhouse Gases to the Atmosphere," *BioScience* 50 (2000): 766. For an overview of recent debates on this topic, see Philip M. Fearnside and Salvador Pueyo, "Greenhouse-Gas Emissions from Tropical Dams," *Nature Climate Change* 2 (2012): 382–84; Andreas Maeck et al.,

"Sediment Trapping by Dams Creates Methane Emission Hot Spots," *Environmental Science & Technology* 47 (2013): 8130–37; and Raquel Mendonça et al., "Greenhouse Gas Emissions from Hydroelectric Reservoirs: What Knowledge Do We Have and What Is Lacking?," in *Greenhouse Gases—Emission, Measurement and Management*, ed. Guozhiang Liu (InTech, 2012). It stands to reason that emissions from reservoirs within China will be vitally important, yet there is at present virtually no research being carried out in this regard. See Yuanan Hu and Hefa Cheng, "The Urgency of Assessing the Greenhouse Gas Budgets of Hydroelectric Reservoirs in China," *Nature Climate Change* 3 (2013): 708–12.

59. For a measured discussion of the lack of scientific consensus on this issue, see Joel Avruch Goldenfum, "Challenges and Solutions for Assessing the Impact of Freshwater Reservoirs on Natural GHG Emissions," *Ecohydrology & Hydrobiology* 12 (2012): 115–22. To signal one example of our lack of knowledge, recent research suggests that river surfaces downstream of large reservoirs in the tropics may also be a significant source of carbon and methane emissions via degasification processes. See Frédéric Guérin et al., "Methane and Carbon Dioxide Emissions from Tropical Reservoirs: Significance of Downstream Rivers," *Geophysical Research Letters* 33 (2006): L21407.

60. See Bakker and Bridge, "Material Worlds?"

61. For critical assessments of IWRM, see François Molle, "Nirvana Concepts, Narratives and Policy Models: Insights from the Water Sector," *Water Alternatives* 1 (2008): 132–36.

62. See James H. Thorp, Martin C. Thoms, and Michael D. Delong, "The Riverine Ecosystem Synthesis: Biocomplexity Across Space and Time," *River Research and Applications* 22 (2006): 123–47.

63. See Poff et al., "Homogenization," 5732. See also William L. Graf, "Downstream Hydrologic and Geomorphic Effects of Large Dams on American Rivers," *Geomorphology* 79 (2006): 336–60.

64. For a useful summary of these approaches, see N. Leroy Poff et al., "The Natural Flow Regime," *BioScience* (1997): 780–82.

CHAPTER SEVEN

1. Stamm, "Bureau of Reclamation International Technical Assistance," 20.

2. In Richard Norgaard's evocative and prescient phrasing, the alteration of rivers is a product of the coevolution of multiple processes too often viewed by development planners, scientists, and the general public as singular and linear. See Richard Norgaard, *Development Betrayed: The End of Progress and a Coevolutionary Revisioning of the Future* (London: Routledge, 1994), 23–48. Norgaard's masterful history serves as a major inspiration for the present work due to both its interdisciplinary breadth and its normative vision.

3. See Sheila Jasanoff, "The Idiom of Co-production," in *States of Knowledge: The Co-Production of Science and Social Order*, ed. Sheila Jasanoff (London: Routledge, 2004), 1–12.

4. I am here thinking of the seminal contributions of scholars of critical development studies—Timothy Mitchell, James Ferguson, Anna Tsing, and Tania Murray Li being the most prominent—that have been featured as theoretical touchstones throughout this book. These thinkers and numerous others have done invaluable service in tracing the multiple genealogies of development interventions throughout the twentieth century as set down, instigated, transformed, and resisted in specific places.

5. For excellent overviews of recent debates within geography and a conceptual explication of spatial scale that aligns with the thinking presented here, see Gordon Macleod

and Martin Jones, "Territorial, Scalar, Networked, Connected: In What Sense a 'Regional World?'" *Regional Studies* 41 (2007): 1177–91. In a vein that resonates with how I interpret spatial scale throughout this book, Moore perceives "scalar categorizations"—the "construction, reproduction and dissolution of specific scalar classifications, and the various social practices involved" in their usages—as more useful than more stringent notions of geographical scale. See Adam Moore, "Rethinking Scale as a Geographical Category: From Analysis to Practice," *Progress in Human Geography* 32 (2008): 215.

6. Technological achievement became particularly important to American policy makers during the Cold War, when modernization theorists "were convinced that the technological edge of the United States would prove decisive in its competition with communist rivals to offer a model of development for the nations of the 'Third World.'" See Adas, *Dominance by Design*, 10. This idea of an American "technological edge" in global affairs certainly predates the Cold War era. It also enabled, for example, the modernizing impulse in water resource development in the early decades of the twentieth century. Ironically, it was the power of this faith in technology and planning as an almost foolproof way of becoming industrialized, and hence modern, that was "so apparent in American intellectual life in the 1930s" and contributed to the creation of the TVA and Hoover Dam. See Engerman, "Romance of Economic Development," 52.

7. Subfields within the discipline of geography—and here I am thinking of political ecology, resource geography, and similar approaches—have also witnessed a burgeoning and vibrant set of debates regarding how to think about nature, technology, and society not as a relatively bounded set of separate spheres of research and action, but as a highly entangled, hybrid set of concepts and practices. For one particularly relevant example built around an ontological reconsideration of water, see Jamie Linton, *What Is Water?: The History of a Modern Abstraction* (Vancouver: University of British Columbia Press, 2010).

8. Some refer to such constructions as "socionatural networks" while others employ "technonatures." See Bakker and Bridge, "Material Worlds?"; and Damian White and Chris Wilbert, "Introduction: Technonatural Time-Spaces," *Science as Culture* 15 (2006): 95–104.

9. See chapter 1 and Mitchell, *Rule of Experts*, 42.

10. Ferguson, *Anti-Politics Machine*, 18. Ferguson derives his interpretative framework from Foucault's work on the genealogy of the prison in Western society. See Michel Foucault, *Discipline and Punish: The Birth of the Prison* (New York: Random House, 1977).

11. See Tsing, *Friction*, 3–6.

12. Dean Rusk, "Remarks of the Secretary of State at the Closing Session of the Water for Peace Conference," Washington, DC, May 31, 1967, 3. See Floyd E. Dominy papers (1934–2002), box 19, AHC (Laramie).

13. Ibid., 5–6.

14. As of November 2014 the US Congress was considering legislation, endorsed by the Senate Committee on Energy and Natural Resources with bipartisan Democratic and Republican support, that would begin implementation of the plan including removal of four hydroelectric dams. See Jeff Barnard, "Oregon Senator Pushes Klamath Dams Bill," *Eureka Times Standard*, November 13, 2014, http://www.times-standard.com/general-news/20141113/oregon-senator-pushes-klamath-dams-bill.

15. See Hannah Gosnell and Erin Clover Kelly, "Peace on the River?: Social-Ecological Restoration and Large Dam Removal in the Klamath Basin, USA," *Water Alternatives* 3 (2010): 361. Gosnell and Kelly expertly chronicle the manifold details of the case and set it within the context of broader river restoration efforts.

16. While dam removal as a historical trend remains in its infancy relative to the ongoing era of dam construction, communities, environmental organizations, and government agencies at several levels increasingly perceive the removal of dams—of various sizes—as a viable means of restoring, repairing, or otherwise intervening in altered and impaired river systems. Most removals to date have occurred on rivers in North America and Europe and have involved older dams that have outlived their original function or present substantial safety hazards (or both). For recent and insightful examinations of the phenomenon, see Angela T. Bednarek, "Undamming Rivers: A Review of the Ecological Impacts of Dam Removal," *Environmental Management* 27 (2001): 803–14; Sara E. Johnson and Brian E. Graber, "Enlisting the Social Sciences in Decisions about Dam Removal," *BioScience* 52 (2002): 731–38; and Dolly Jørgensen and Birgitta Malm Renöfält, "Damned If You Do, Dammed If You Don't: Debates on Dam Removal in the Swedish Media," *Ecology and Society* 18 (2013): 18.

APPENDIX

1. Bureau consulting teams ranged widely throughout Europe and pockets of European diaspora societies throughout the twentieth century. One of the longest-lasting engagements was in Australia, where over a ten-year period (1951–1961) a cadre of Bureau consultants advised the Australian government on all facets of the Snowy Mountains hydroelectric scheme. The Snowy Mountains project, a massive transmountain water diversion plan, was modeled directly on the Bureau's experience with the Colorado–Big Thompson project in the American West. See Stamm, "Bureau of Reclamation International Technical Assistance," 6.

2. Actually, the State Department and Bureau did not bother with regions, but simply classified most countries according to their common names.

3. For an excellent historical overview, see Kate B. Showers, "Congo River's Grand Inga Hydroelectricity Scheme: Linking Environmental History, Policy and Impact," *Water History* 1 (2009): 31–58.

4. "Summation of the Inga Hydroelectric Project," airgram from American Embassy, Kinshasa, to State Department, December 17, 1970, RG 59, Subject-Numeric Files (1970–1973), box 984, NARA (College Park).

5. See Showers, "Grand Inga," 47–53.

6. The seminal work on this topic is Hart, *Volta River Project.*

7. US Bureau of Reclamation, *Evaluations, Suggestions and Recommendations on the Program of the Volta River Authority for Resettlement of the People in the Area to Be Inundated by Construction of the Volta River Dam, Ghana* (Washington, DC: US Government Printing Office, 1963).

8. William C. Trimble, Memorandum to the Under Secretary, April 4, 1967, RG 59, Subject-Numeric Files (1967–1969), box 455, NARA (College Park).

9. Hendrik van Oss, Memorandum to Joseph Palmer, April 4, 1967, RG 59, Subject-Numeric Files (1967–1969), box 455, NARA (College Park).

10. Trimble, Memorandum, 3.

11. American Embassy (Abidjan) to Secretary of State, April 6, 1967, RG 59, Subject-Numeric Files (1967–1969), box 512, NARA (College Park). See also chap. 1, note 14.

12. Export-Import Bank of Washington, letter to Nicholas de B. Katzenbach, Under Secretary of State, May 8, 1967, RG 59, Subject-Numeric Files (1967–1969), box 512, NARA (College Park).

13. Memorandum of Conversation, January 10, 1969, RG 59, Subject-Numeric Files (1967–1969), box 512, NARA (College Park).

14. See Véronique Lassailly-Jacob, "Land-Based Strategies in Dam-Related Resettlement Programmes in Africa," in *Understanding Impoverishment: The Consequences of Development-Induced Displacement*, ed. Christopher McDowell (Oxford: Berghahn, 1996), 188.

15. US Bureau of Reclamation, *Kano Plain Project, Kenya—Reconnaissance Examination* (Washington, DC: US Bureau of Reclamation, 1967).

16. These details are found in a letter from R. E. Radford of Great Britain's Ministry of Overseas Development to Edmond C. Hutchinson of USAID, January 25, 1966, RG 59, Subject-Numeric File (1964–1966), box 646, NARA (College Park).

17. R. Williams, Memorandum to James Rives, "Division Budgets for Calendar Year 1952," January 14, 1952, RG 469.6, Records of the ICA (1951–1961), General Correspondence, box 18, NARA (College Park).

18. US Bureau of Reclamation, *St. Paul River Project, Liberia—Reconnaissance Report* (Washington, DC: US Bureau of Reclamation, 1952).

19. Incoming Telegram to Secretary of State Office from American Embassy, Monrovia, October 29, 1965, RG 59, Subject-Numeric Files (1964–1966), box 569, NARA (College Park).

20. Wade Williams, "Finally Lighting Liberia: Mt. Coffee Hydro Plant to Be Completed by 2016 in US$65M Pact," FrontPageAfrica, January 25, 2013, http://www.frontpage africaonline.com/politics/42-politics/5140-finally-lighting-liberia-mt-coffee-hydro-to -be-completed-by-2016-in-us65m-pact.html.

21. US Bureau of Reclamation, *Reconnaissance Study—Land and Water Resources of the Lake Chad Basin* (Washington, DC: US Bureau of Reclamation, 1968).

22. George Dolgin, Memorandum to Department of State, "Status of Niger Dam Project," March 16, 1963, RG 59, Subject-Numeric Files (1963), box 44, NARA (College Park).

23. Bruce G. Davis and Howard F. Haworth, *Review of US AID Water Resources Development Project, Somali Republic* (Mogadiscio, Somali Republic: US Agency for International Development, November 1963).

24. US Bureau of Reclamation, *Reconnaissance Appraisal—Land and Water Resource Development Plans and Potentials, Rufiji River Basin* (Washington, DC: US Bureau of Reclamation, March 1967), 2–3.

25. See Hoag, "Transplanting the TVA?" Complementary to my analysis of the Pa Mong Dam on the Mekong River (see chap. 5), Hoag documents the "undoing" of the Stiegler's Gorge Dam on the Rufiji River, which was never built, focusing on the shifting internal political dynamics that gave rise to opposition to the project.

26. Nick Cullather's exegesis of the Helmand Valley project is immensely important in drawing these pieces together. See Cullather, "Damming Afghanistan."

27. Ernie Hood, "Putting a New Face on Old Afghanistan Is One of Morrison-Knudsen's Projects," *Engineering News-Record*, February 21, 1952, 45–49.

28. Cullather, "Damming Afghanistan."

29. Dominy, oral history interviews, 76–77.

30. T. Mundal (Chief Engineer, International Engineering Company), letter to R. R. Will (Country Director, USOM Pakistan), January 7, 1955, RG 469.6, Records of the ICA (1951–1961), General Correspondence, box 487, NARA (College Park).

31. R. R. Will, Telegram to Secretary of State, January 11, 1955, RG 469.6, Records of the ICA (1951–1961), General Correspondence, box 487, NARA (College Park).

32. T. Mundal, letter to R. R. Will, Karachi, January 31, 1955, RG 469.6, Records of the ICA (1951–1961), General Correspondence, box 487, NARA (College Park).

33. Said Hasan, letter to James Baird, March 23, 1955, RG 469.6, Records of the ICA (1951–1961), General Correspondence, box 487, NARA (College Park).

34. See Saila Parveen and I. M. Faisal, "People versus Power: The Geopolitics of Kaptai Dam in Bangladesh," *International Journal of Water Resources Development* 18 (2002): 197–208.

35. Recent scholarship has revealed the extent to which India's monumental dam-building and river basin development programs were modeled after the US experience, especially the TVA. See Klingensmith, "One Valley."

36. John L. Savage, letter to Michael Straus, Assistant Secretary of the Interior, May 26, 1945, RG 115, Yangtze Gorge Project (1943–1949), box 10, NARA (Denver).

37. See US Bureau of Reclamation, *Evaluation of Engineering and Economic Feasibility—Beas and Rajasthan Project, Northern India* (Washington, DC: US Bureau of Reclamation, July 1963); and US Bureau of Reclamation, *Beas Dam/Beas Project, Talwara, Punjab, India* (Washington, DC: US Bureau of Reclamation, January 1965).

38. I use "Korea" throughout this section to reflect the place name used in project reports and documents, a usage that remained common despite the official designation of "South Korea" in 1948. The retention of "Korea" in US reports is itself a geopolitical statement.

39. Korean Hydroelectric Power Feasibility Survey Group, *Report on Reconnaissance of Kum Gang Multiple-Purpose Project, Chungju-Yoju Hydroelectric Project, Hongchon-Chunchon Hydroelectric Project, Imkei Hydroelectric Project in the Republic of Korea* (Seoul, June 1950).

40. John C. Coyne, "Staff Study—Republic of Korea Power System," July 28, 1955, RG 115, General Correspondence (1946–1960), box 5, FRC, NARA (Denver).

41. United Nations Economic Commission for Asia and the Far East (ECAFE), *Multiple-Purpose River Basin Development, part 2D, Water Resources Development in Afghanistan, Iran, Republic of Korea and Nepal* (Bangkok: ECAFE, 1961), 57.

42. US Bureau of Reclamation and Geological Survey, *Reconnaissance Report/Water Resources Study—Han River Basin* (Washington, DC: US Bureau of Reclamation, 1971).

43. US Bureau of Reclamation and Geological Survey, *Han River Basin, Republic of Korea—Preliminary Survey* (Washington, DC: US Bureau of Reclamation, June 1965), 54.

44. US Bureau of Reclamation, *Annual Report 1975* (Denver: Bureau of Reclamation, 1975).

45. See UNESCO, *Facing the Challenges: United Nations World Water Development Report 3, Case Studies Volume* (Paris: UNESCO, 2009), 33–35.

46. For a detailed summary of US foreign policy goals in Laos during the 1950s, see Karabell, *Architects of Intervention*, 206–24.

47. Hiroshi Hori, *The Mekong: Environment and Development* (Tokyo: United Nations University Press, 2000), 160–61.

48. See Nguyen, *Mekong River*, 123–27.

49. Rutherford Poats, Memorandum to Jack Valenti, "Lower Mekong Basin Development Scheme," March 26, 1965, DDRS, http://galenet.galegroup.com/servlet/DDRS (Doc. No. CK3100409056).

50. Willard A. Hanna, *The Test at Nam Ngum*, part 4 of *The Mekong Project, AUFS Southeast Asia Series*, vol. 16, no. 13 (Hanover, NH: American Universities Field Staff, 1968), 1.

51. Hanna, *Test at Nam Ngum*, 2.

52. Nguyen, *Mekong River*, 171.

53. See James Wescoat, Sarah Halvorson, and Daanish Mustafa, "Water Management in the Indus Basin of Pakistan: A Half-Century Perspective," *Water Resources Development* 16 (2000): 391–406.

54. See Kenneth F. Vernon, transcript of Bureau of Reclamation oral history interviews conducted and edited by Brit Allan Storey, April 1995, in Fullerton, CA, NARA (College Park), 344–78.

55. Wescoat, Halvorson, and Mustafa, "Water Management," 393–95.

56. Robert W. Kennedy, "White House Considers Economic Aid to Pakistan for the Tarbela Dam," May 28, 1963, DDRS, http://galenet.galegroup.com/servlet/DDRS (Doc. No. CK3100218610).

57. Robert Komer, "Robert Komer Discusses George Ball's Mission to Pakistan to Reassure Ayub That the U.S. Will Back Them If They Are Attacked by India," August 12, 1963, DDRS, http://galenet.galegroup.com/servlet/DDRS (Doc. No. CK3100327649). See also "Instructions for Under Secretary Ball's Mission to Pakistan," August 28, 1963, DDRS, http://galenet.galegroup.com/servlet/DDRS (Doc. No. CK3100284527).

58. It has also produced a number of negative social and ecological consequences never anticipated during the project's planning stages. See Asianics Agro-Development International (Pvt) Ltd., *Tarbela Dam and Related Aspects of the Indus River Basin, Pakistan (Report)* (Cape Town: World Commission in Dams, 2000), http://www.dams.org/docs/kbase/studies/cspkmain.pdf.

59. I. Baird, Telegram from the Foreign Service of the United States of America, January 31, 1955, RG 469.6, Records of the ICA (1951–1961), General Correspondence, box 810, NARA (College Park).

60. See Michael Cernea, "Public Policy Responses to Development-Induced Population Displacements," *Economic and Political Weekly* (1996): 1517.

61. Albert Ravenholt, *Hydroelectric Power and Philippine Industrialization*, AUFS Reports, Southeast Asia Series, vol. 1 (1953) & vol. 2 (1954) (New York: American Universities Field Staff, July 10, 1953), 87–93.

62. D. R. Burnett, *NEC-AID Philippine Water Resources Survey, First Semi-Annual Report* (Manila: US Bureau of Reclamation, January 1964), 2.

63. US Bureau of Reclamation, *A Report on the Central Luzon Basin, Luzon Island, Philippines* (Manila: US Bureau of Reclamation, November 1966), i.

64. Burnett, *NEC-AID Philippine Water Resources Survey*, 44.

65. William I. Palmer, "A Review of Bureau of Reclamation Investigations of 7 Major River Basins in the Philippines," August 17 through 27, 1965, RG 115, International Reports (1914–1987), box 9, NARA (Denver).

66. US Bureau of Reclamation, *Annual Report 1975.*

67. US Bureau of Reclamation, *A Report on the Cagayan River Basin, Luzon Island, Philippines* (Manila: US Bureau of Reclamation, December 1966), 37–41.

68. See Charles Drucker, "Dam the Chico: Hydro Development and Tribal Resistance in the Philippines," in *Social and Environmental Effects of Large Dams*, vol. 2: *Case Studies*, ed. Goldsmith and Hildyard, 305–9.

69. "American Activity in Ceylon Embraces Expanding Sphere," *Christian Science Monitor*, August 9, 1949, 10.

70. "Point Four Aiding Peace, Prosperity," *New York Times*, October 15, 1949, 22.

71. Quoted in Clifford MacFadden, "The Gal Oya Valley: Ceylon's Little TVA," *Geographical Review* 44 (1954): 273.

72. Ibid., 279.

73. US Bureau of Reclamation, *Wu-Sheh Dam: Final Design and Construction Report* (Washington, DC: US Bureau of Reclamation, April 1960), 2–7.

74. Yi Sun, "Militant Diplomacy: The Taiwan Strait Crises and Sino-American Relations, 1954–1958," in *Eisenhower Administration*, ed. Statler and Johns, 130.

75. Karl M. Rankin, "Report on Foreign Economic Policy Discussions between U.S. Officials in the Far East and Clarence B. Randall and Associates: Taiwan," December 1, 1956, 1–3, DDRS, http://galenet.galegroup.com/servlet/DDRS (Doc. No. CK3100279735).

76. A. K. Hamilton, letter to Wilbur A. Dexheimer (Commissioner, USBR), July 23, 1955, RG 115, General Correspondence (1946–1960), box 31, FRC, NARA (Denver).

77. US Bureau of Reclamation, *Shihmen Reservoir Project, Formosa (Review of a Definite Plan Report prepared by the Shihmen Planning Commission)* (Washington, DC: US Bureau of Reclamation, October 1955), 26.

78. US Bureau of Reclamation, *Tseng-wen Reservoir Project, Taiwan* (Washington, DC: US Bureau of Reclamation, September 1964).

79. Richard J. Shukle, *Advisory Report on Water Resources Policy, Administration and Development for the Council for International Economic Cooperation and Development, Executive Yuan, Republic of China* (Washington, DC: US Bureau of Reclamation, March 1965), v–vi.

80. William J. Sheppard, "Report on Foreign Economic Policy Discussions between U.S. Officials in the Far East and Clarence B. Randall and Associates: Thailand," December 1, 1956, 1, DDRS, http://galenet.galegroup.com/servlet/DDRS (Doc. No. CK3100279735).

81. For a theoretically rich and empirically detailed account of Thai-US relations and their implications, see James Glassman, *Thailand at the Margins: Internationalization of the State and the Transformation of Labour* (Oxford: Oxford University Press, 2004).

82. US Bureau of Reclamation, *Yanhee Project, Thailand: Power, Irrigation, Flood Control, and Navigation*, vol. 1, *Project Evaluation* (Washington, DC: US Bureau of Reclamation, December 1955), i.

83. E. R. Dexter et al. (Bureau Team in Thailand), Memorandum to Chief Engineer, "Mission to Thailand—Investigation of Yan Hee Project," December 21, 1953, RG 115, General Correspondence (1946–1960), box 32, FRC, NARA (Denver).

84. Memorandum from the Director of the Office of Southeast Asian Affairs (Young) to the Assistant Secretary of State for Far Eastern Affairs (Robertson), June 7, 1956; in FRUS 1955–1957, vol. 22, Southeast Asia (Washington, DC: GPO, 1954), 888–89.

85. Telegram from the Department of State to the Embassy in Thailand, September 29, 1956; in *FRUS, 1955–1957*, vol. 22, 903–4.

86. R. P. Lightfoot, "Problems of Resettlement in the Development of River Basins in Thailand," in *River Basin Planning: Theory and Practice*, ed. Suranjit K. Saha and Christopher J. Barrow (Chichester, UK: John Wiley and Son, 1981), 94.

87. US Bureau of Reclamation, Survey of Water Resources, *Mun and Chi River Basins, Northeastern Thailand* (Washington, DC: US Bureau of Reclamation, March 1965), i.

88. See Peter F. Bell, "Thailand's Northeast: Regional Underdevelopment, 'Insurgency,' and Official Response," *Pacific Affairs* 42 (1969); and Charles Keyes, "Hegemony and Resistance in Northeastern Thailand," in *Regions and National Integration in Thailand, 1892–1992*, ed. Volker Grabowsky (Wiesbaden: Harrassowitz Verlag, 1995), 154–82.

89. See Sneddon, "Reconfiguring Scale and Power," 2238–41.

90. Anthony J. Perry, *Report on Field Trip to Iran* (Washington, DC: US Department of the Interior, Bureau of Reclamation, August 1952), 32–33.

91. "Iraq Resumes Plans to Tap Water Resources," *Engineering News-Record*, July 28, 1960, 49.

92. U. V. Engstrom, "Trip Reports Summarizing Activities of U. V. Engstrom in Iraq during the Period May 7, 1954 to August 4, 1954," RG 115, International Reports (1914–1987), box 4, NARA (Denver).

93. Perry, *Report*, 25. Perry carried out his assignment while "on loan" to the State Department from the Bureau under what was called a "Point 4 assignment," an arrangement typical of the Bureau's early forays into international activities.

94. US Bureau of Reclamation, *Karaj River Project, Iran: Evaluation Report* (Washington, DC: US Bureau of Reclamation, May 1954), 3.

95. For a much more detailed analysis of this period, see Steve Marsh, "Continuity and Change: Reinterpreting the Policies of the Truman and Eisenhower Administration toward Iran, 1950–1954," *Journal of Cold War Studies* 7 (2005): 79–123.

96. Operations Coordinating Board (National Security Council), "Progress Report on NSC 5402, United States Policy towards Iran," September 28, 1954, 1, DDRS, http://galenet.galegroup.com/servlet/DDRS (Doc. No. CK3100263984).

97. Ibid., 9.

98. A much more detailed account of the Karaj Dam can be found in Cyrus Schayegh, "Iran's Karaj Dam Affair: Emerging Mass Consumerism, the Politics of Promise, and the Cold War in the Third World," *Comparative Studies in Society and History* 54 (2012): 633–38.

99. See Tucker, "Containing Communism," 148–51; and Ekbladh, "'Mr. TVA,'" 373.

100. American Embassy Tehran, November 4, 1964, 4, RG 59, Subject-Numeric Files (1964–1966), box 1660, NARA (College Park).

101. US Bureau of Reclamation, *Proposed Yarmouk-Jordan Valley Project and Minor Wadis* (Washington, DC: US Bureau of Reclamation, 1953).

102. Memorandum from American Embassy (Amman, Jordan) to State Department, Khalid Bin al Walid Dam. May 17, 1967, RG 59, Subject-Numeric Files (1967–1969), box 98, NARA (College Park).

103. Ibid.

104. Philip R. Dickinson, *Reclamation Advisory Team to Turkey—Annual History, Calendar Year 1955* (Ankara, Turkey: Bureau of Reclamation, December 31, 1955), 1.

105. Ibid., 6.

106. See Leila M. Harris, "Water and Conflict Geographies of the Southeastern Anatolia Project," *Society & Natural Resources* 15 (2002): 743–59.

107. Marshall T. Jones and Charles H. Ducote, "Hydroelectric Power Development in Argentina," *Commerce Reports, November 24, 1930* (Washington, DC: Department of Commerce, 1930), 492.

108. Thomas L. Hughes, "Soviets Evince Interest in Joint Aid Venture with British Firm in Argentina," Intelligence Note to Secretary of State, Washington, DC, October 16, 1967, RG 59, Subject-Numeric Files (1967–1969), box 986, NARA (College Park).

109. American Embassy (Buenos Aires), "Bidding Procedures on Salto Grande Hydro-Electric Project," Telegram to Department of State, Washington, DC, January 24, 1973, RG 59, Subject-Numeric Files (1970–1973), box 981, NARA (College Park).

110. American Embassy (Buenos Aires), "Salto Grande Engineering Contract Award," Telegram to Department of State, Washington, DC, January 26, 1973, RG 59, Subject-Numeric Files (1970–1973), box 981, NARA (College Park).

111. American Embassy (Buenos Aires), "Salto Grande Project Award to Charles T. Main," Telegram to Department of State, Washington, DC, May 21, 1973, RG 59, Subject-Numeric Files (1970–1973), box 981, NARA (College Park).

112. Arthur F. Johnson, *Villa Montes Irrigation Project—Tarija Province, Bolivia* (Denver: Bureau of Reclamation, October 1962), RG 115, International Reports (1914–1987), box 2, NARA (Denver).

113. For Paulo Affonso, see "Great Power Plant Is Opened by Brazil," *New York Times*, January 16, 1955, 23. For the Três Marias project, see A. N. Rydland, "Brazil's Tres Marias Project Puts Rio Sao Francisco to Work," *Engineering News-Record*, August 11, 1960, 42–44.

114. US Bureau of Reclamation, *Reconnaissance Appraisal—Land and Water Resources, Arguaia-Tacontins River Basin, Brazil* (Washington, DC: US Bureau of Reclamation, December 1964), RG 115, International Reports (1914–1987), box 10, NARA (Denver).

115. Ibid., 3.

116. US Bureau of Reclamation, *Reconnaissance Appraisal—Land and Water Resource: Rio São Francisco Basin, Brazil* (Washington, DC: US Bureau of Reclamation, 1967), 1–2, RG 115, International Reports (1914–1987), box 10, NARA (Denver).

117. "Brazilian Request for Soviet Investments," Central Intelligence Agency Information Report, September 14, 1962, RG 59, Central Decimal Files (1960–1963), box 32, NARA (College Park).

118. T. P. Ahrens, "Elqui River Project and Other Work in Chile," letter to Assistant Commissioner and Chief Engineer, May 7, 1954, RG 115, NARA (Denver).

119. US Bureau of Reclamation, *Reconnaissance Appraisal—Land and Water Resources: Magdalena-Cauca River Basin, Colombia* (Washington, DC: US Bureau of Reclamation, April 1967). The "look" of the Bureau reports produced during the 1960s in Latin America and elsewhere, and indeed, much of the language used in describing the development potential of various river basins, is quite similar. One suspects that once a model for reconnaissance studies in the context of foreign operations was established, the Bureau authors of later studies borrowed heavily from these pioneering works. This observation attests to my contention that the Bureau, up to a certain point, was operating under a universalizing assumption regarding the basic features and transformability of all river systems.

120. Ibid., 7.

121. American Embassy (Bogota), "Subject: BECM-Chivor Project," letter to State Department, Washington, DC, December 28, 1972, RG 59, Subject-Numeric Files (1970–1973), box 985, NARA (College Park).

122. US Bureau of Reclamation, *Progress Report on Tempisque Valley Project Investigations, Guanacaste Province, Costa Rica* (Washington, DC: US Bureau of Reclamation, July 1952), 2–3.

123. "Costa Rican Hydroelectric Dam Is Seen as Progressing on Schedule," *New York Times*, July 30, 1956, 30.

124. American Embassy (San Jose), Telegram to Department of State, "Boruca Hydroelectric Project for Aluminum Smelter," May 31, 1973, RG 59, Subject-Numeric Files (1970–1973), box 986, NARA (College Park). The still unfinished Boruca Dam, originally envisioned as a supplier of electricity to a large aluminum smelting operation run by Alcoa, remains a priority of the government and has generated substantial opposition due to its likely impacts on indigenous people and ecosystems in the vicinity of the dam site. See Elizabeth P. Anderson, Catherine M. Pringle, and Manrique Rojas, "Transforming Tropical Rivers: An Environmental Perspective on Hydropower Development in Costa Rica," *Aquatic Conservation: Marine and Freshwater Ecosystems* 16 (2006): 679–93.

125. Dominic S. Pastir, *Report on Development of the Yaque del Norte River Basin (Dominican Republic): Comprehensive Water Resources Development Analysis* (Washington, DC: US Bureau of Reclamation, September 1974).

126. American Embassy (Santo Domingo), "National Movement for the Construction of the Tavera Dam," Memorandum to the Department of State, January 13, 1968, RG 59, Subject-Numeric Files (1967–1969), box 1972, NARA (College Park). For a contemporaneous perspective on the nationalist fervor surrounding the Tavera Dam, see Nancie Gonzalez, "The Sociology of a Dam," *Human Organization* 31 (1972): 353–60. Gonzalez concludes that despite widespread support for the project from peasants and other marginalized groups throughout Dominican society, the dam's primary beneficiaries would be the landowning class and urban elites.

127. Peter Phillips, F. Arturo Russell, and John Turner, "Effect of Non-Point Source Runoff and Urban Sewage on Yaque del Norte River in Dominican Republic," *International Journal of Environment and Pollution* 31 (2007): 249.

128. Raoul St. Lô, Marcel Thebaud, and George Hargreaves, "The Artibonite Valley Project in Haiti for Irrigation, Drainage, Flood Control and Hydroelectric Power Development" (Washington, DC, 1956), 4–5, RG 115, General Correspondence (1946–1960), box 5, FRC, NARA (Denver).

129. Philip Howard, "Development-Induced Displacement in Haiti," *Refuge: Canada's Journal on Refugees* 16 (September 1997): 6.

130. Benson Timmons III, Telegram from the Embassy in Haiti to the Department of State (Port-au-Prince), October 21, 1965, *FRUS, 1964–1968*, vol. 32, *Dominican Republic; Cuba; Haiti; Guyana* (Washington, DC: GPO, 1965), 805.

131. Howard, "Development-Induced Displacement," 6–10.

132. "IDB Makes $20 Million Grant for Haiti Hydro," December 15, 2011, www.iadb.org/en/news/news-releases.

133. George A. Fleming, *Report on the Feasibility Survey of Hydroelectric Development Possibilities in Honduras, C. A., Point IV Project No. (1) 1461211* (Washington, DC: US Bureau of Reclamation, June 26, 1951), 2–4.

134. Robinson, *Water for the West*, 91. See also Philippus Wester, Edwin Rap, and Sergio Vargas-Velazquez, "The Hydraulic Mission and the Mexican Hydrocracy: Regulating and Reforming the Flows of Water and Power," *Water Alternatives* 2 (2009): 399.

135. Robert J. Newell, *Reconnaissance Report: Nicaraguan Power Investigation Mission, Managua, Nicaragua* (Washington, DC: US Bureau of Reclamation, July 1951), 1–2.

136. Ibid., 57–59.

137. See Kolko, *Confronting the Third World*, 284–87; and Westad, *Global Cold War*, 339–48.

138. Memorandum from Acting Assistant Commissioner to Chief Engineer re: Nicaraguan Point Four Mission, March 7, 1951, RG 115, International Reports (1914–1987), box 8, NARA (Denver).

139. US Bureau of Reclamation, *Hacia La Meta: Evaluation Report (Prepared for the Government of Nicaragua)* (Washington, DC: US Bureau of Reclamation, 1977), 43.

140. Ibid., 4.

141. Ibid., 5.

142. "Nicaragua set to break ground on 253-MW Tumarin hydroelectric project," *Hydro World.com*, January 29, 2014, http://www.hydroworld.com/articles/2014/01/nicaragua-set-to-break-ground-on-253-mw-tumarin-hydroelectric-project.html.

143. US Bureau of Reclamation, *Tinajones Project* (Washington, DC: US Bureau of Reclamation, August 1961); and US Bureau of Reclamation, *Tumbes Project* (Washington, DC: US Bureau of Reclamation, September 1961).

144. Bryan R. Frisbie, "Peru's Decision Regarding the Mantaro Project," Memorandum to Department of State, March 23, 1966, RG 59, Subject-Numeric Files (1964–1966), box 2775, NARA (College Park).

145. US Bureau of Reclamation, *Report on Proposed Plan of Development of Rincon de Baygorria Hydroelectric Site on Rio Negro, Uruguay* (Washington, DC: US Bureau of Reclamation, March 1954). The dam, along with one additional hydroelectric scheme, was completed over several phases in the 1960s and 1970s.

146. Richard D. Harding, "Russian Professor Indicates Soviet Interest in Salto Grande Hydroelectric Project," Memorandum to Department of State, November 7, 1964, RG 59, Subject-Numeric Files (1964–1966), box 1964, NARA (College Park). These and similar statements can probably be interpreted as examples of institutional paranoia

on the part of American officials as well as savvy rhetoric on the part of Soviet representatives.

147. Richard B. Owen, "Problems in Guri Dam Construction," Memorandum to Department of State, April 19, 1966, RG 59, Subject-Numeric Files (1967–1969), box 3061, NARA (College Park).

148. American Embassy (Caracas), "The Guri Dam Is Inaugurated," Memorandum to Department of State, November 13, 1968, RG 59, Subject-Numeric Files (1967–1969), box 1972, NARA (College Park). The guest list—which included representatives of American businesses, a host of government agencies, and academic institutions such as Harvard and MIT—is instructive in terms of the powerful actors assembled in the service of large dams during the Cold War era.

BIBLIOGRAPHY

Archive and Manuscript Collections

This book is based on archival research carried out over several periods between 2008 and 2012. The archives I used included planning documents, reports, studies, technical investigations, newspaper clippings, and inter-agency communications (of a variety of classifications) of the Bureau of Reclamation (Record Group 115), primarily from the period 1945–1975. These materials are stored at the National Archives and Research Administration (NARA) Rocky Mountain Region offices in Denver, Colorado. I also investigated similar archival materials in NARA's Denver office that are awaiting NARA classification and available upon special request, but stored under the rubric of the Federal Records Center (FRC). I accessed similar communiqués and documents in State Department (Record Group 59) archives, as well as those of the United States Agency for International Development and its predecessors (Record Groups 286 and 469, respectively), housed at NARA's College Park, Maryland, facility. I also used the *Foreign Relations of the United States* (*FRUS*) series as well as the Declassified Documents Reference System (DDRS), an online storehouse of over 70,000 formerly classified documents related to various presidential administrations. Finally, I explored the collected papers of former Bureau of Reclamation engineers at the American Heritage Center at the University of Wyoming in Laramie. To my knowledge, many, if not most, of the documents regarding the Bureau's international activities have not been previously employed for public research.

SOURCES OF ARCHIVAL MATERIALS
American Heritage Center, Laramie, Wyoming
    Floyd F. Dominy papers, 1934–1970
    Robert F. Herdman papers, 1927–1966

National Archives and Research Administration, College Park, Maryland
    Record Group 59: General Records of the US Department of State
    Record Group 286: Records of the Agency for International Development
    Record Group 469: Records of US Foreign Assistance Agencies, 1948–1961
National Archives and Research Administration, Denver, Colorado
    Federal Records Center
    Record Group 115: Records of the US Bureau of Reclamation

PERIODICALS

Asahi Shimbun (Japan)
Asia Times Online
Christian Science Monitor
Commerce Reports
Daily Star (Lebanon)
Engineering News-Record
Ethiopia Observer
Eureka Times Standard
FrontPageAfrica
Horn Affairs
Irrawaddy
Nation
New Republic
News (Pakistan)
New York Times
Voice of America
Washington Daily News
Washington Post
Washington Post Times Herald

REPORTS OF THE BUREAU OF RECLAMATION AND AFFILIATED REPORTS

Burnett, D. R. *NEC-AID Philippine Water Resources Survey, First Semi-Annual Report*. Manila: Bureau of Reclamation, January 1964.

Davis, Bruce G., and Howard F. Haworth. *Review of US AID Water Resources Development Project, Somali Republic*. Mogadiscio, Somali Republic: US Agency for International Development, November 1963.

Dickinson, Philip R. *Reclamation Advisory Team to Turkey—Annual History, Calendar Year 1955*. Ankara, Turkey: Bureau of Reclamation, December 31, 1955.

Fleming, George A. *Report on the Feasibility Survey of Hydroelectric Development Possibilities in Honduras, C. A., Point IV Project No. (1) 1461211*. Washington, DC: US Bureau of Reclamation, June 26, 1951.

Johnson, Arthur F. *Villa Montes Irrigation Project—Tarija Province, Bolivia*. Denver, CO: Bureau of Reclamation, October 1962.

Korean Hydroelectric Power Feasibility Survey Group. *Report on Reconnaissance of Kum Gang Multiple-Purpose Project, Chungju-Yoju Hydroelectric Project, Hongchon-Chunchon Hydroelectric Project, Imkei Hydroelectric Project in the Republic of Korea*. Seoul, June 1950.

Newell, Robert J. *Reconnaissance Report: Nicaraguan Power Investigation Mission, Managua, Nicaragua*. Washington, DC: US Bureau of Reclamation, July 1951.

Pastir, Dominic S. *Report on Development of the Yaque del Norte River Basin (Dominican Republic): Comprehensive Water Resources Development Analysis.* Washington, DC: US Bureau of Reclamation, September 1974.

Perry, Anthony J. *Report on Field Trip to Iran.* Washington, DC: US Bureau of Reclamation, August 1952.

Shukle, Richard J. *Advisory Report on Water Resources Policy, Administration and Development for the Council for International Economic Cooperation and Development, Executive Yuan, Republic of China.* Washington, DC: US Bureau of Reclamation, March 1965.

United States (US) Bureau of Reclamation. *Reconnaissance Report on the Litani River Project, Lebanon.* Washington, DC: Department of the Interior, US Bureau of Reclamation, June 1951.

———. *St. Paul River Project, Liberia—Reconnaissance Report.* Washington, DC: US Bureau of Reclamation, 1952.

———. *Progress Report on Tempisque Valley Project Investigations, Guanacaste Province, Costa Rica.* Washington, DC: US Bureau of Reclamation, July 1952.

———. *Proposed Yarmouk-Jordan Valley Project and Minor Wadis.* Washington, DC: US Bureau of Reclamation, 1953.

———. *Report on Proposed Plan of Development of Rincon de Baygorria Hydroelectric Site on Rio Negro, Uruguay.* Washington, DC: US Bureau of Reclamation, March 1954.

———. *Development Plan for the Litani River Basin, Republic of Lebanon.* Vol. 1, *General Description and Economic Analysis.* Beirut: Litani River Investigation Staff, 1954.

———. *Karaj River Project, Iran: Evaluation Report.* Washington, DC: US Bureau of Reclamation, May 1954.

———. *Shihmen Reservoir Project, Formosa (Review of a Definite Plan Report prepared by the Shihmen Planning Commission).* Washington, DC: US Bureau of Reclamation, October 1955.

———. *Yanhee Project, Thailand: Power, Irrigation, Flood Control, and Navigation.* Vol. 1, *Project Evaluation.* Washington, DC: US Bureau of Reclamation, December 1955.

———. *Reconnaissance Report: Lower Mekong River Basin.* Prepared for International Cooperation Administration. US Bureau of Reclamation, March 1956.

———. *Water Resources Investigations for the Nahr El-Barad Basin, Reconnaissance Report.* Denver, CO: US Bureau of Reclamation, September 1956.

———. *Wu-Sheh Dam: Final Design and Construction Report.* Washington, DC: US Bureau of Reclamation, April 1960.

———. *Tinajones Project.* Washington, DC: US Bureau of Reclamation, August 1961.

———. *Tumbes Project.* Washington, DC: US Bureau of Reclamation, September 1961.

———. *Ethiopia-United States Cooperative Program for the Study of Water Resources—1961 Annual Report.* Washington, DC: US Bureau of Reclamation, October 31, 1961.

———. *Preliminary Study, Pa Mong Project, Thailand-Laos.* Denver, CO: US Bureau of Reclamation, January 1962.

———. *Evaluations, Suggestions and Recommendations on the Program of the Volta River Authority for Resettlement of the People in the Area to be Inundated by Construction of the Volta River Dam, Ghana.* Washington, DC: US Government Printing Office, 1963.

———. *Evaluation of Engineering and Economic Feasibility—Beas and Rajasthan Project, Northern India.* Washington, DC: US Bureau of Reclamation, July 1963.

———. *Land and Water Resources of the Blue Nile Basin, Ethiopia.* Washington, DC: US Bureau of Reclamation, 1964.

———. *Tseng-wen Reservoir Project, Taiwan.* Washington, DC: US Bureau of Reclamation, September 1964.

———. *Reconnaissance Appraisal—Land and Water Resources, Arguaia-Tacontins River Basin, Brazil.* Washington, DC: US Bureau of Reclamation, December 1964.

———. *Beas Dam/Beas Project, Talwara, Punjab, India.* Washington, DC: US Bureau of Reclamation, January 1965.

———. *Survey of Water Resources, Mun and Chi River Basins, Northeastern Thailand.* Washington, DC: US Bureau of Reclamation, March 1965.

———. *Mekong Pa Mong Survey (Phase I): Interim Report.* Denver, CO: US Bureau of Reclamation, June 1965.

———. *Pa Mong Project (Lower Mekong River Basin), Phase I Report,* volume 2, appendix I *(Agreements).* Denver, CO: US Bureau of Reclamation, March 1966.

———. *A Report on the Central Luzon Basin, Luzon Island, Philippines.* Manila: US Bureau of Reclamation, November 1966.

———. *A Report on the Cagayan River Basin, Luzon Island, Philippines.* Manila: US Bureau of Reclamation, December 1966.

———. *Kano Plain Project, Kenya—Reconnaissance Examination.* Washington, DC: US Bureau of Reclamation, 1967.

———. *Reconnaissance Appraisal—Land and Water Resource Development Plans and Potentials, Rufiji River Basin.* Washington, DC: US Bureau of Reclamation, March 1967.

———. *Reconnaissance Appraisal—Land and Water Resource: Rio São Francisco Basin, Brazil.* Washington, DC: US Bureau of Reclamation, 1967.

———. *Reconnaissance Appraisal—Land and Water Resources: Magdalena-Cauca River Basin, Colombia.* Washington, DC: US Bureau of Reclamation, April 1967.

———. *Reconnaissance Study—Land and Water Resources of the Lake Chad Basin.* Washington, DC: US Bureau of Reclamation, 1968.

———. *Pa Mong Project, Stage One, Phase II (Executive Summary).* Denver, CO: US Bureau of Reclamation, 1970.

———. *Pa Mong Phase II—Appendix 5, Plans and Estimates.* Denver, CO: United States Bureau of Reclamation, 1972.

———. *Annual Report 1975.* Denver, CO: US Bureau of Reclamation, 1975.

———. *Hacia La Meta: Evaluation Report* (prepared for the Government of Nicaragua). Washington, DC: US Bureau of Reclamation, 1977, 43.

US Bureau of Reclamation and Geological Survey. *Han River Basin, Republic of Korea—Preliminary Survey.* Washington, DC: US Bureau of Reclamation, June 1965.

US Bureau of Reclamation and Geological Survey. *Reconnaissance Report/Water Resources Study—Han River Basin.* Washington, DC: US Bureau of Reclamation, 1971.

PUBLISHED PRIMARY AND SECONDARY SOURCES

Adams, William. *Wasting the Rain: Rivers, People and Planning in Africa.* Minneapolis: University of Minnesota Press, 1992.

Adamson, Michael R. "'The Most Important Single Aspect of Our Foreign Policy'? The Eisenhower Administration, Foreign Aid, and the Third World." In *The Eisenhower Administration, the Third World, and the Globalization of the Cold War,* edited by Kathryn C. Statler and Andrew L. Johns, 47–72. Lanham, MD: Rowman & Littlefield, 2006.

Adas, Michael. *Dominance by Design: Technological Imperatives and America's Civilizing Mission.* Cambridge, MA: Belknap Press of Harvard University Press, 2006.

Agnew, John. *Hegemony: The New Shape of Global Power.* Philadelphia: Temple University Press, 2005.

Akera, Atsushi. "Constructing a Representation for an Ecology of Knowledge: Methodological Advances in the Integration of Knowledge and Its Various Contexts." *Social Studies of Science* 37 (2007): 413–41.

Alatout, Samer. "Bringing Abundance into Environmental Politics: Constructing a Zionist Network of Water Abundance, Immigration, and Colonization." *Social Studies of Science* 39 (2009): 363–94.

———. "'States' of Scarcity: Water, Space, and Identity Politics in Israel, 1948–1959." *Environment and Planning D: Society and Space* 26 (2008): 959–82.

Allen, John. *Lost Geographies of Power.* Malden, MA: Blackwell, 2003.

———. "Powerful Assemblages?" *Area* 43 (2011): 154–57.

Anderson, Ben, and Colin McFarlane. "Assemblage and Geography." *Area* 43 (2011): 124–27.

Anderson, Elizabeth P., Catherine M. Pringle, and Manrique Rojas. "Transforming Tropical Rivers: An Environmental Perspective on Hydropower Development in Costa Rica." *Aquatic Conservation: Marine and Freshwater Ecosystems* 16 (2006): 679–93.

Asdal, Kristin, Brita Brenna, and Ingunn Moser. "Re-inventing Politics of the State: Science and the Politics of Contestation." In *Technoscience: The Politics of Interventions,* edited by Kristin Asdal, Brita Brenna, and Ingunn Moser, 7–53. Oslo: Oslo Academic Press, 2007.

Awulachew, Seleshi B., Matthew McCartney, Tammo S. Steenhuis, and Abdalla A. Ahmed. *A Review of Hydrology Sediment and Water Resource Use in the Blue Nile Basin.* IWMI Working Paper 131. Colombo, Sri Lanka: International Water Management Institute, 2008.

Bakker, Karen. "The Politics of Hydropower: Developing the Mekong." *Political Geography* 18 (1999): 209–32.

Bakker, Karen, and Gavin Bridge. "Material Worlds? Resource Geographies and the 'Matter of Nature.'" *Progress in Human Geography* 30 (2006): 5–27.

Barnes, Jessica. "Managing the Waters of Ba'th Country: The Politics of Water Scarcity in Syria." *Geopolitics* 14 (2009): 510–30.

Barrow, Christopher J. "River Basin Development and Management: A Critical Review." *World Development* 26 (1998): 171–86.

Bednarek, Angela T. "Undamming Rivers: A Review of the Ecological Impacts of Dam Removal." *Environmental Management* 27 (2001): 803–14.

Bell, Peter F. "Thailand's Northeast: Regional Underdevelopment, 'Insurgency,' and Official Response." *Pacific Affairs* 42 (1969): 47–54.

Berger, Mark. *The Battle for Asia: From Decolonization to Globalization.* London: RoutledgeCurzon, 2004.

Bevir, Mark. "What Is Genealogy?" *Journal of the Philosophy of History* 2 (2008): 263–75.

Biggs, David. "Reclamation Nations: The U.S. Bureau of Reclamation's Role in Water Management and Nation-Building in the Mekong Valley, 1945–1975." *Comparative Technology Transfer and Society* 4 (2006): 225–46.

Bijker, Wiebe. "Dikes and Dams, Thick with Politics." *Isis* 98 (2007): 109–23.

Billington, David B., and Donald C. Jackson. *Big Dams of the New Deal Era: A Confluence of Engineering and Politics.* Norman: University of Oklahoma Press, 2006.

Birkenholz, Trevor. "Groundwater Governmentality: Hegemony and Technologies of Resistance in Rajasthan's (India) Groundwater Governance." *Geographical Journal* 175 (2009): 208–20.

Black, Brian. "Organic Planning: The Intersection of Nature and Economic Planning in the Early Tennessee Valley Authority." *Journal of Environmental Policy & Planning* 4 (2002): 157–68.

Block, Paul J., Kenneth Strzepek, and Bajali Rajagopalan. *Integrated Management of the Blue Nile Basin in Ethiopia: Hydropower and Irrigation Modeling*. Washington, DC: International Food Policy and Research Institute, May 2007.

Bochinski, Feliks, and William Diamond. "TVA's in the Middle East." *Middle East Journal* 4 (1950): 52–82.

Boxer, Baruch. "China's Three Gorges Dam: Questions and Prospects." *China Quarterly* 113 (1988): 94–108.

Braun, Bruce. "Environmental Issues: Global Natures in the Space of Assemblage." *Progress in Human Geography* 30 (2006): 644–54.

Brautigan, Deborah. "Aid with 'Chinese Characteristics': Chinese Foreign Aid and Development Finance Meet the OECD-DAC Regime." *Journal of International Development* 23 (2011): 752–64.

Briscoe, John. "The Changing Face of Water Infrastructure Financing in Developing Countries." *International Journal of Water Resources Development* 15 (1999): 301–8.

———. "Overreach and Response: The Politics of the WCD and Its Aftermath." *Water Alternatives* 3 (2010): 399–415.

Carmody, Padraig, Godfrey Hampwaye, and Enock Sakala. "Globalisation and the Role of the State? Chinese Geogovernance in Zambia." *New Political Economy* 17 (2012): 209–29.

Castree, Noel. "False Antitheses? Marxism, Nature and Actor-Networks." *Antipode* 34 (2002): 111–46.

Cernea, Michael. "Public Policy Responses to Development-Induced Population Displacements." *Economic and Political Weekly* (1996): 1515–23.

Chandler, William U. *The Myth of the TVA: Conservation and Development in the Tennessee Valley, 1933–1980*. Cambridge, MA: Ballinger, 1984.

Chao, Benjamin Fong. "Anthropogenic Impact on Global Geodynamics Due to Reservoir Water Impoundment." *Geophysical Research Letters* 22 (1995): 3529–32.

Chetham, Deirdre. *Before the Deluge: The Vanishing World of the Yangtze's Three Gorges*. New York: Palgrave Macmillan, 2002.

Clapp, Gordon. *The TVA: An Approach to the Development of a Region*. Chicago: University of Chicago Press, 1955.

Collier, Stephen J., and Aihwa Ong. "Global Assemblages: Anthropological Problems." In *Global Assemblages: Technology, Politics, and Ethics as Anthropological Problems*, edited by Aihwa Ong and Stephen J. Collier, 3–21. Malden, MA: Blackwell, 2005.

Comair, Fady. "Litani Water Management: Prospect for the Future." Paper presented at the Congres International de Kaslik–Liban, June 18–20, 1998. http://www.samana.funredes.org/agua/files/geopolitique/COMIER.rtf.

Conway, Declan. "A Water Balance Model of the Upper Blue Nile in Ethiopia." *Hydrological Sciences Journal* 42 (1997): 265–86.

Cox, Robert. "Gramsci, Hegemony and International Relations: An Essay in Method." *Millennium: Journal of International Studies* 12 (1983): 162–75.

Crow, Ben. *Sharing the Ganges: The Politics and Technology of River Development*. Thousand Oaks, CA: Sage Publications, 1995.

Cullather, Nick. "Damming Afghanistan: Modernization in a Buffer State." *Journal of American History* 89 (2002): 512–37.

———. *The Hungry World: America's Cold War Battle against Poverty in Asia*. Cambridge, MA: Harvard University Press, 2010.

Dai Qing. *Yangtze! Yangtze!* Edited by P. Adams and J. Thibodeau. London: Probe International, 1994.

de Laet, Marianne, and Annemarie Mol. "The Zimbabwe Bush Pump: Mechanics of a Fluid Technology." *Social Studies of Science* 30 (2002): 225–63.

DeLanda, Manuel. *A New Philosophy of Society: Assemblage Theory and Social Complexity.* London: Continuum, 2006.

Dirlik, Arif. "Confounding Metaphors, Inventions of the World: What Is World History For?" In *Writing World History, 1800–2000*, edited by Benedikt Stuchtey and Eckhardt Fuchs, 91–133. London: Oxford University Press, 2003.

Dodds, Klaus. *Global Geopolitics: A Critical Introduction.* Essex, UK: Prentice Hall, 2005.

Doel, Ronald E. "Constituting the Postwar Earth Sciences: The Military's Influence on the Environmental Sciences in the USA after 1945." *Social Studies of Science* 33 (2003): 635–66.

Dore, John, Yu Xiaogang, and K. Yuk-shing Li. "China's Energy Reforms and Hydropower Expansion in Yunnan." In *Democratizing Water Governance in the Mekong Region*, edited by Louis Lebel, John Dore, Rajesh Daniel, and Yang Saing Koma, 55–92. Chiang Mai, Thailand: Mekong Press, 2007.

Drucker, Charles. "Dam the Chico: Hydro Development and Tribal Resistance in the Philippines." In Goldsmith and Hildyard, *Social and Environmental Effects of Large Dams*, vol. 2, *Case Studies*, 301–18.

Dubash, Navroz. "Reflections on the WCD as a Mechanism of Global Governance." *Water Alternatives* 3 (2010): 416–22.

Dudziak, Mary. *Cold War Civil Rights: Race and the Image of American Democracy.* Princeton, NJ: Princeton University Press, 2000.

Ekbladh, David. *The Great American Mission: Modernization and the Construction of an American World Order.* Princeton, NJ: Princeton University Press, 2010.

———. "'Mr. TVA': Grass-Roots Development, David Lilienthal, and the Rise and Fall of the Tennessee Valley Authority as a Symbol for U.S. Overseas Development, 1933–1973." *Diplomatic History* 26 (2002): 335–74.

Ekers, Michael, and Alex Loftus. "The Power of Water: Developing Dialogues between Foucault and Gramsci." *Environment and Planning D: Society and Space* 26 (2008): 698–718.

Ellison, Karin. "Explaining Hoover, Grand Coulee, and Shasta Dams: Institutional Stability and Professional Identity in the U.S. Bureau of Reclamation." In *Reclamation, Managing Water in the West: The Bureau of Reclamation: History Essays from the Centennial Symposium*, vol. 1, edited by US Department of Interior, Bureau of Reclamation, 221–48. Denver, CO: US Department of Interior, Bureau of Reclamation, 2008.

Engerman, David. "The Romance of Economic Development and New Histories of the Cold War." *Diplomatic History* 28 (2004): 23–54.

Farvar, M. Taghi, and John P. Milton, eds. *The Careless Technology: Ecology and International Development.* Garden City, NY: Natural History Press, 1972.

Fearnside, Philip M. "China's Three Gorges Dam: 'Fatal' Project or Step towards Modernization?" *World Development* 16 (1988): 615–30.

Fearnside, Philip M., and Salvador Pueyo. "Greenhouse-Gas Emissions from Tropical Dams." *Nature Climate Change* 2 (2012): 382–84.

Feis, Herbert. *The China Tangle: The American Effort in China from Pearl Harbor to the Marshall Mission.* Princeton, NJ: Princeton University Press, 1953.

Ferguson, James. *The Anti-Politics Machine: "Development," Depoliticization, and Bureaucratic Power in Lesotho.* Minneapolis: University of Minnesota Press, 1994.

Finer, Matt, and Clinton N. Jenkins. "Proliferation of Hydroelectric Dams in the Andean Amazon and Implications for Andes-Amazon Connectivity." *PLoS ONE* 7 (2012): e35126. doi: 10.1371/jounal.pone.35126.

Fisher, William, ed. *Toward Sustainable Development: Struggling over India's Narmada River.* London: M. E. Sharpe, 1995.

Fletcher, Robert. "When Environmental Issues Collide: Climate Change and the Shifting Political Ecology of Hydroelectric Power." *Peace and Conflict Monitor* 5 (2011): 14–30.

Flyvbjerg, Bent. "Five Misunderstandings about Case-Study Research." *Qualitative Inquiry* 12 (2006): 219–45.

Foreign Relations of the United States: Diplomatic Papers 1943. Vol. 4, *The Near East and Africa.* Washington, DC: GPO, 1943.

Foreign Relations of the United States: Diplomatic Papers, 1945. Vol. 7, *The Far East, China.* Washington, DC: GPO, 1945.

Foreign Relations of the United States, 1952–1954. Vol. 9, part 1, *The Near and Middle East.* Washington, DC: GPO, 1953.

Foreign Relations of the United States, 1952–1954. Vol. 12, part 1, *East Asia and the Pacific.* Washington, DC: GPO, 1954.

Foreign Relations of the United States, 1955–1957. Vol. 14, *Arab-Israeli Dispute,* 1955. Washington, DC: GPO, 1955.

Foreign Relations of the United States, 1955–1957. Vol. 22, *Southeast Asia.* Washington, DC: GPO, 1954.

Foreign Relations of the United States, 1964–1968. Vol. 32, *Dominican Republic; Cuba; Haiti; Guyana.* Washington, DC: GPO, 1965.

Foucault, Michel. *Discipline and Punish: The Birth of the Prison.* New York: Random House, 1977.

Frost, Stephen. "Chinese Outward Direct Investment in Southeast Asia: How Big are the Flows and What Does It Mean for the Region?" *Pacific Review* 17 (2004): 323–40.

Gaddis, John Lewis. *Strategies of Containment: A Critical Appraisal of American National Security Policy during the Cold War.* 2nd ed. New York: Oxford University Press, 2005.

Gilman, Nils. *Mandarins of the Future: Modernization Theory in Cold War America.* Baltimore, MD: Johns Hopkins University Press, 2003.

Glassman, James. "From Seattle (and Ubon) to Bangkok: The Scales of Resistance to Corporate Globalization." *Environment and Planning D: Society and Space* 19 (2001): 513–33.

———. *Thailand at the Margins: Internationalization of the State and the Transformation of Labour.* Oxford: Oxford University Press, 2004.

Goldenfum, Joel Avruch. "Challenges and Solutions for Assessing the Impact of Freshwater Reservoirs on Natural GHG Emissions." *Ecohydrology & Hydrobiology* 12 (2012): 115–22.

Goldman, Michael. *Imperial Nature: The World Bank and Struggles for Justice in the Age of Globalization.* New Haven, CT: Yale University Press, 2005.

Goldsmith, Edward, and Nicholas Hildyard. *Social and Environmental Effects of Large Dams.* Vol. 1, *Overview.* San Francisco: Sierra Club Books, 1984.

———, eds. *Social and Environmental Effects of Large Dams.* Vol. 2, *Case Studies.* Cornwall, UK: Wadebridge Ecological Centre, 1986.

Golze, Alfred R. "Multipurpose Investigation of the Blue Nile." *Civil Engineering* 696 (1959): 36–40.

Gonzalez, Nancie. "The Sociology of a Dam." *Human Organization* 31 (1972): 353–60.

Gosnell, Hannah, and Erin Clover Kelly. "Peace on the River? Social-Ecological Restoration and Large Dam Removal in the Klamath Basin, USA." *Water Alternatives* 3 (2010): 361–83.

Graf, William L. "Dam Nation: A Geographic Census of American Dams and Their Large-scale Hydrological Impacts." *Water Resources Research* 35 (1999): 1305–11.

———. "Downstream Hydrologic and Geomorphic Effects of Large Dams on American Rivers." *Geomorphology* 79 (2006): 336–60.

Gramsci, Antonio. *Selections from the Prison Notebooks*. New York: International Publishers, 1971.

Grumbine, R. Edward, and Maharaj K. Pandit, "Threats From India's Himalaya Dams." *Science* 339 (January 4, 2013): 36–37.

Grumbine, R. Edward, and Jianchu Xu. "Mekong Hydropower Development." *Science* 332 (2011): 178–79.

Grumbine, R. Edward, John Dore, and Jianchu Xu. "Mekong Hydropower: Drivers of Change and Governance Challenges." *Frontiers in Ecology and the Environment* 10 (2012): 91–98.

Guérin, Frédéric, Gwenaël Abril, Sandrine Richard, Benoît Burban, Cécile Reynouard, Patrick Seyler, and Robert Delmas. "Methane and Carbon Dioxide Emissions from Tropical Reservoirs: Significance of Downstream Rivers." *Geophysical Research Letters* 33 (2006): L21407.

Hammond, Michael. "The Grand Ethiopian Renaissance Dam and the Blue Nile: Implications for Transboundary Water Governance." GWF Discussion Paper 1307. Canberra: Global Water Forum, 2013. http://www.globalwaterforum.org/2013/02/18/the-grand-ethiopian-renaissance-dam-and-the-blue-nile-implications-for-transboundary-water-governance/.

Hanna, Willard A. *The Test at Nam Ngum*. Part 4 of *The Mekong Project*. AUFS Southeast Asia Series, vol. 16, no. 13. Hanover, NH: American Universities Field Staff, 1968.

———. *The Prize at Pa Mong*. Part 5 of *The Mekong Project*. AUFS Southeast Asia Series, vol. 16, no. 14. Hanover, NH: American Universities Field Staff, 1968.

Harding, Sandra. "After the Neutrality Debate: Science, Politics, and 'Strong Objectivity.'" *Social Research* 59 (1992): 567–87.

Harris, Leila M. "Water and Conflict Geographies of the Southeastern Anatolia Project." *Society &Natural Resources* 15 (2002): 743–59.

Hart, David. *The Volta River Project: A Case Study in Politics and Technology*. Edinburgh: Edinburgh University Press, 1980.

Hayes, Samuel P. "Point Four in United States Foreign Policy." *Annals of the American Academy of Political and Social Science* 268 (1950): 27–35.

Hecht, Gabrielle, ed. *Entangled Geographies: Empire and Technopolitics in the Global Cold War*. Cambridge, MA: MIT Press, 2011.

———. "Introduction." In *Entangled Geographies: Empire and Technopolitics in the Global Cold War*, edited by Gabrielle Hecht, 1–12. Cambridge, MA: MIT Press, 2011.

Hensengerth, Oliver. "Chinese Hydropower Companies and Environmental Norms in Countries of the Global South: The Involvement of Sinohydro in Ghana's Bui Dam." *Environment, Development and Sustainability* 15 (2013): 285–300.

Hirsch, Philip. "The Changing Political Dynamics of Dam Building on the Mekong." *Water Alternatives* 3 (2010): 312–23.

———. "China and the Cascading Geopolitics of Lower Mekong Dams." *Asia-Pacific Journal* 9 (2011): 1–4.

———. "Water Governance Reform and Catchment Management in the Mekong Region." *Journal of Environment & Development* 15 (2006): 184–201.

Hoag, Heather. *Developing the Rivers of East and West Africa: An Environmental History*. London: Bloomsbury, 2013.

———. "Transplanting the TVA? International Contributions to Postwar River Development in Tanzania." *Comparative Technology Transfer and Society* 4 (2006): 247–68.

Hobsbawm, Eric. *The Age of Extremes: A History of the World, 1914–1991*. New York: Vintage Books, 1994.

Holder, Curtis. "Contested Visions: Technology Transfer, Water Resources, and Social Capital in Chilascó, Guatemala." *Comparative Technology Transfer and Society* 4 (2006) 269–86.

Hori, Hiroshi. *The Mekong: Environment and Development.* Tokyo: United Nations University Press, 2000.

Howard, Philip. "Development-Induced Displacement in Haiti." *Refuge: Canada's Journal on Refugees* 16 (September 1997): 4–11.

Hu, Yuanan, and Hefa Cheng. "The Urgency of Assessing the Greenhouse Gas Budgets of Hydroelectric Reservoirs in China." *Nature Climate Change* 3 (2013): 708–12.

Hudson, James. "The Litani River of Lebanon: An Example of Middle Eastern Water Development." *Middle East Journal* 25 (1971): 1–14.

Hughes, J. Donald. *What Is Environmental History?* Cambridge: Polity Press, 2006.

Hughes, Thomas. *Networks of Power: Electrification in Western Society, 1880–1930.* Baltimore, MD: Johns Hopkins University Press, 1983.

Hyndman, Jennifer. "Mind the Gap: Bridging Feminist and Political Geography through Geopolitics." *Political Geography* 23 (2004): 307–22.

Iggers, Georg. *Historiography in the Twentieth Century: From Scientific Objectivity to the Postmodern Challenge.* Middletown, CT: Wesleyan University Press, 1997.

International Commission on Large Dams (ICOLD). *Dams and the World's Water.* Paris: ICOLD, 2007.

International Rivers. *The New Great Walls: A Guide to China's Overseas Dam Industry.* 2nd ed. Berkeley, CA: International Rivers, November 2012.

Intracaso, David M. "The Politics of Technology: The 'Unpleasant Truth about Pleasant Dam.'" *Western Historical Quarterly* 26 (1995): 333–52.

Isaacman, Allen E., and Barbara S. Isaacman. *Dams, Displacement, and the Delusion of Development: Cahora Bassa and Its Legacies in Mozambique.* Athens: Ohio University Press, 2013.

Isaacman, Allen, and Chris Sneddon. "Toward a Social and Environmental History of the Building of Cahora Bassa Dam." *Journal of Southern African Studies* 26 (2000): 597–632.

Jackson, Donald C. "Engineering in the Progressive Era: A New Look at Frederick Haynes Newell and the U.S. Reclamation Service." *Technology and Culture* 34 (1993): 539–74.

Jasanoff, Sheila. "The Idiom of Co-production." In *States of Knowledge: The Co-Production of Science and Social Order,* edited by Sheila Jasanoff, 1–12. London: Routledge, 2004.

Johnson, Lyndon B. "Johnson on Southeast Asian Aid." *Current History* 49 (1966): 303–8.

Johnson, Sara E., and Brian E. Graber. "Enlisting the Social Sciences in Decisions about Dam Removal." *BioScience* 52 (2002): 731–38.

Jones, William C., and Marsha Freeman. "Three Gorges Dam: The TVA on the Yangtze River." *21st Century* (Fall 2000): 24–46.

Jørgensen, Dolly, and Birgitta Malm Renöfält. "Damned If You Do, Dammed If You Don't: Debates on Dam Removal in the Swedish Media." *Ecology and Society* 18 (2013): 18.

Josephson, Paul R. *Industrialized Nature: Brute Force Technology and the Transformation of the Natural World.* Washington, DC: Island Press, 2002.

———. "'Projects of the Century' in Soviet History: Large-Scale Technologies from Lenin to Gorbachev." *Technology and Culture* 36 (1995): 519–59.

Kapuściński, Ryszard. *The Emperor: Downfall of an Autocrat.* New York: Vintage Books, 1983.

Karabell, Zachary. *Architects of Intervention: The United States, the Third World, and the Cold War, 1946–1962.* Baton Rouge: Louisiana State University Press, 1999.

Keyes, Charles. "Hegemony and Resistance in Northeastern Thailand." In *Regions and National Integration in Thailand, 1892–1992,* edited by Volker Grabowsky, 154–82. Wiesbaden: Harrassowitz Verlag, 1995.

Khagram, Sanjeev. *Dams and Development: Transnational Struggles for Water and Power*. Ithaca, NY: Cornell University Press, 2004.

Kilby, Patrick. "The Changing Development Landscape in the First Decade of the 21st Century and Its Implications for Developing Countries." *Third World Quarterly* 33 (2012): 1001–17.

Kirsch, Scott, and Don Mitchell. "The Nature of Things: Dead Labor, Nonhuman Actors and the Persistence of Marxism." *Antipode* 36 (2004): 687–705.

Klingensmith, David. *"One Valley and a Thousand": Dams, Nationalism, and Development*. New Delhi: Oxford University Press, 2007.

Kolko, Gabriel. *Another Century of War?* New York: The New Press, 2002.

———. *Confronting the Third World: United States Foreign Policy, 1945–1980*. New York: Pantheon Books, 1988.

Kopinski, Dominik, Andrzej Polus, and Ian Taylor. "Contextualising Chinese Engagement in Africa." *Journal of Contemporary African Studies* 29 (2011): 129–36.

Koppes, Clayton R. "Efficiency, Equity, Esthetics: Shifting Themes in American Conservation." In *The Ends of the Earth: Perspectives on Modern Environmental History*, edited by Donald Worster, 230–51. Cambridge: Cambridge University Press, 1988.

Kragelund, Peter. "Knocking on a Wide Open Door: Chinese Investments in Africa." *Review of African Political Economy* 36 (2009): 479–97.

Kuus, Merje. "Professionals of Geopolitics: Agency in International Politics." *Geography Compass* 2 (2008): 2062–79.

LaBounty, J. F. "Assessment of the Environmental Effects of Constructing the Three Gorge Project on the Yangtze River." *Water International* 9 (1984): 10–17.

Larner, Wendy, and William Walters. "The Political Rationality of 'New Regionalism': Toward a Genealogy of the Region." *Theory and Society* 31 (2002): 391–432.

Laron, Guy. *Origins of the Suez Crisis: Postwar Development Diplomacy and the Struggle over Third World Industrialization, 1945–1956*. Baltimore, MD: Johns Hopkins University Press, 2013.

Lassailly-Jacob, Véronique. "Land-Based Strategies in Dam-Related Resettlement Programmes in Africa." In *Understanding Impoverishment: The Consequences of Development-Induced Displacement*, edited by Christopher McDowell, 187–99. Oxford: Berghahn, 1996.

Latham, Michael E. *Modernization as Ideology: American Social Science and "Nation Building" in the Kennedy Era*. Chapel Hill: University of North Carolina Press, 2000.

Latour, Bruno. "Drawing Things Together." In *Representation in Scientific Practice*, edited by Michael Lynch and Steven Woolgar, 19–68. Cambridge, MA: MIT Press, 1990.

———. *Reassembling the Social: An Introduction to Actor-Network-Theory*. Oxford: Oxford University Press, 2005.

Li, Tania Murray. *The Will to Improve: Governmentality, Development, and the Practice of Politics*. Durham, NC: Duke University Press, 2007.

Lightfoot, R. P. "Problems of Resettlement in the Development of River Basins in Thailand." In *River Basin Planning: Theory and Practice*, edited by Suranjit K. Saha and Christopher J. Barrow, 93–114. Chichester, UK: John Wiley and Son, 1981.

Linton, Jamie. *What Is Water? The History of a Modern Abstraction*. Vancouver: University of British Columbia Press, 2010.

Little, Douglas. "His Finest Hour? Eisenhower, Lebanon, and the 1958 Middle East Crisis." In *Empire and Revolution: The United States and the Third World since 1945*, edited by Peter L. Hahn and Mary Ann Heiss, 17–47. Columbus: Ohio State University Press, 2001.

MacFadden, Clifford. "The Gal Oya Valley: Ceylon's Little TVA." *Geographical Review* 44 (1954): 271–81.

Macleod, Gordon, and Martin Jones. "Territorial, Scalar, Networked, Connected: In What Sense a 'Regional World'?" *Regional Studies* 41 (2007): 1177–91.

Maeck, Andreas, Tonya DelSontro, Daniel F. McGinnis, Helmut Fischer, Sabine Flury, Mark Schmidt, Peer Fietzek, and Andreas Lorke. "Sediment Trapping by Dams Creates Methane Emission Hot Spots." *Environmental Science & Technology* 47 (2013): 8130–37.

Magilligan, Francis J., and Keith H. Nislow. "Changes in Hydrologic Regime by Dams." *Geomorphology* 71 (2005): 61–78.

Mahayni, Basil. "Producing Crisis: Hegemonic Debates, Mediations and Representations of Water Scarcity." In *Contemporary Water Governance in the Global South: Scarcity, Marketization and Participation,* edited by Leila Harris, Jacqui Goldin, and Christopher Sneddon, 35–44. New York: Routledge, 2013.

Marcus, Harold G. *A History of Ethiopia.* 2nd ed. Berkeley: University of California Press, 2002.

Markakis, John. *Ethiopia: The Last Two Frontiers.* Rochester, NY: James Currey, 2011.

Marsh, Steve. "Continuity and Change: Reinterpreting the Policies of the Truman and Eisenhower Administration toward Iran, 1950–1954." *Journal of Cold War Studies* 7 (2005): 79–123.

Mawdsley, Emma. "The Changing Geographies of Foreign Aid and Development Cooperation: Contributions from Gift Theory." *Transactions of the Institute of British Geographers* 37 (2012): 256–72.

McCann, James. "Ethiopia, Britain, and the Negotiations for the Lake Tana Dam, 1922–1935." *International Journal of African Historical Studies,* 14 (1981): 667–69.

McCully, Patrick. *Silenced Rivers: The Ecology and Politics of Large Dams.* 2nd ed. London: Zed Books, 2001.

McDonald, Kristen, Peter Bosshard, and Nicole Brewer. "Exporting Dams: China's Hydropower Industry Goes Global." *Journal of Environmental Management* 90 (2009): S294–S302.

McFarlane, Colin, and Ben Anderson. "Thinking with Assemblage." *Area* 43 (2011): 162–64.

McMahon, Robert J. "Introduction: The Challenge of the Third World." In *Empire and Revolution: The United States and the Third World since 1945,* edited by Peter L. Hahn and Mary Ann Heiss, 1–14. Columbus: Ohio State University Press, 2001.

McNeill, J. R. "Observations on the Nature and Culture of Environmental History." *History and Theory* 42 (2003): 5–43.

McNeill, J. R., and Corinna Unger, eds. *Environmental Histories of the Cold War.* Cambridge: Cambridge University Press, 2010.

McPhee, John. *Encounters with the Archdruid.* New York: Farrar, Straus and Giroux, 1977.

Mekong Committee. *Brief Description of the Pa Mong Project.* Bangkok: Economic Commission for Asia and the Far East, June 30, 1961.

Mendonça, Raquel, Nathan Barros, Luciana O. Vidal, Felipe Pacheco, Sarian Kosten, and Fábio Roland. "Greenhouse Gas Emissions from Hydroelectric Reservoirs: What Knowledge Do We Have and What Is Lacking?" In *Greenhouse Gases—Emission, Measurement and Management,* ed. Guozhiang Liu. InTech, 2012.

Mitchell, Timothy. *Rule of Experts: Egypt, Techno-Politics, Modernity.* Berkeley: University of California Press, 2002.

Molle, François. "Nirvana Concepts, Narratives and Policy Models: Insights from the Water Sector." *Water Alternatives* 1 (2008): 131–56.

———. "River-Basin Planning and Management: The Social Life of a Concept." *Geoforum* 40 (2009): 484–94.

Moore, Adam. "Rethinking Scale as a Geographical Category: From Analysis to Practice." *Progress in Human Geography* 32 (2008): 203–25.

Moore, Deborah, John Dore, and Dipak Gyawali. "The World Commission on Dams +10: Revisiting the Large Dams Controversy." *Water Alternatives* 3 (2010): 3–13.

Murdoch, Jonathan. "Ecologising Sociology: Actor-Network Theory, Co-construction and the Problem of Human Exemptionalism." *Sociology* 35 (2001): 111–33.

Murdoch, Jonathan, and Terry Marsden. "The Spatialization of Politics: Local and National Actor-Spaces in Environmental Conflict." *Transactions of the Institute of British Geographers*, 20 (1995): 368–80.

Müller, Martin. "Opening the Black Box of the Organization: Socio-Material Practices of Geopolitical Ordering." *Political Geography* 31 (2013): 379–88.

———. "Reconsidering the Concept of Discourse for the Field of Critical Geopolitics: Towards Discourse as Language *and* Practice." *Political Geography* 27 (2008): 322–38.

Nega, Tsegaye H. "Saving Wild Rice: The Rise and Fall of Nett Lake Dam." *Environment and History* 14 (2008): 5–39.

Neumann, Roderick P. *Making Political Ecology.* London: Hodder Arnold, 2005.

Nguyen, Thi Dieu. *The Mekong River and the Struggle for Indochina: Water, War, and Peace.* Westport, CT: Praeger, 1999.

Norgaard, Richard. *Development Betrayed: The End of Progress and a Coevolutionary Revisioning of the Future.* London: Routledge, 1994.

Ogden, Laura, Nik Heynen, Ulrich Oslender, Paige West, Karim-Aly Kassam, and Paul Robbins. "Global Assemblages, Resilience, and Earth Stewardship in the Anthropocene." *Frontiers in Ecology and the Environment* 11 (2013): 341–47.

Oldenziel, Ruth. *Making Technology Masculine: Men, Women and Modern Machines in America, 1870–1945.* Amsterdam: Amsterdam University Press, 1999.

Oliver, John. "The TVA Power Program." *Current History* 34 (1958): 257–64.

O'Neill, Karen M. "Why the TVA Remains Unique: Interest Groups and the Defeat of New Deal River Planning." *Rural Sociology* 67 (2002): 163–82.

Ó Tuathail, Gearóid. "General Introduction: Thinking Critically about Geopolitics." In *The Geopolitics Reader*, 2nd ed., edited by Gearóid Ó Tuathail et al., 1–14. New York: Routledge, 2006.

Ó Tuathail, Gearóid, and John Agnew. "Geopolitics and Discourse: Practical Geopolitical Reasoning in American Foreign Policy." *Political Geography Quarterly* 11 (1992): 190–204.

Papadopoulos, Dimitris. "Alter-Ontologies: Towards a Constituent Politics in Technoscience." *Social Studies of Science* 41 (2011): 177–201.

Parveen, Saila, and I. M. Faisal. "People Versus Power: The Geopolitics of Kaptai Dam in Bangladesh." *International Journal of Water Resources Development* 18 (2002): 197–208.

Pearce, Fred. *The Dammed: Rivers, Dams, and the Coming World Water Crisis.* London: Bodley Head, 1992.

Perkins, John H. *Geopolitics and the Green Revolution: Wheat, Genes, and the Cold War.* New York: Oxford University Press, 1997.

Peyton, Jonathan. "Corporate Ecology: BC Hydro's Stikine-Iskut Project and the Unbuilt Environment." *Journal of Historical Geography* 37 (2011): 358–69.

Phadke, Rupali. "Assessing Water Scarcity and Watershed Development in Maharashtra, India: A Case Study of the Baliraja Memorial Dam." *Science, Technology, & Human Values* 27 (2002): 236–61.

Phillips, Peter, F. Arturo Russell, and John Turner. "Effect of Non-Point Source Runoff and Urban Sewage on Yaque del Norte River in Dominican Republic." *International Journal of Environment and Pollution* 31 (2007): 244–66.

Pisani, Donald J. *Water and American Government: The Reclamation Bureau, National Water Policy, and the West, 1902–1935.* Berkeley: University of California Press, 2002.

Poff, N. LeRoy, J. David Allan, Mark B. Bain, James R. Karr, Karen L. Prestegaard, Brian D. Richter, Richard E. Sparks, and Julie C. Stromberg. "The Natural Flow Regime." *BioScience* (1997): 769–84.

Poff, N. LeRoy, Julian D. Olden, David M. Merritt, and David M. Pepin. "Homogenization of Regional River Dynamics by Dams and Global Biodiversity Implications." *Proceedings of the National Academy of Sciences* 104 (2007): 5732–37.

Power, Marcus, Giles Mohan, and May Tan-Mullins. *China's Resource Diplomacy in Africa: Powering Development?* New York: Palgrave Macmillan, 2012.

Rahman, Muhammad Mizanur, and Olli Varis. "Integrated Water Resources Management: Evolution, Prospects and Future Challenges." *Sustainability: Science, Practice, & Policy* 1 (2005): 15–21.

Raphaeli, Nimrod. "Development Planning: Lebanon." *Western Political Quarterly* 20 (1967): 714–28.

Ravenholt, Albert. *Hydroelectric Power and Philippine Industrialization.* AUFS Reports, Southeast Asia Series, vol. 1 (1953) & vol. 2 (1954), 87–93. New York: American Universities Field Staff, July 10, 1953.

Reisner, Marc. *Cadillac Desert: The American West and Its Disappearing Water.* New York: Viking, 1986.

Reuss, Martin. "Seeing Like an Engineer: Water Projects and the Mediation of the Incommensurable." *Technology and Culture* 49 (2008): 531–46.

Rhodes, B. "From Cooksville to Chungking: The Dam-Designing Career of John L. Savage." *Wisconsin Magazine of History* 72 (Summer 1989): 242–72.

Richter, Brian D., Sandra Postel, Carmen Revenga, Thayer Scudder, Bernhard Lehner, Allegra Churchill, and Morgan Chow. "Lost in Development's Shadow: The Downstream Human Consequences of Dams." *Water Alternatives* 3 (2010): 14–42.

Rist, Gilbert. *The History of Development: From Western Origins to Global Faith.* London: Zed Books, 2002.

Robbins, Paul. *Political Ecology: A Critical Introduction.* 2nd ed. Malden, MA: Wiley-Blackwell, 2012.

Robinson, Michael C. *Water for the West: The Bureau of Reclamation, 1902–1977.* Chicago: Public Works Historical Society, 1979.

Rook, Robert. "Race, Water, and Foreign Policy: The Tennessee Valley Authority's Global Agenda Meets 'Jim Crow.'" *Diplomatic History* 28 (2004): 55–81.

Rowley, William. *The Bureau of Reclamation: Origins and Growth to 1945.* Vol. 1. Denver, CO: Bureau of Reclamation, 2006.

Schaller, Michael. *The United States and China in the Twentieth Century.* 2nd ed. New York: Oxford University Press, 1990.

———. *The U.S. Crusade in China, 1938–1945.* New York: Columbia University Press, 1979.

Schayegh, Cyrus. "Iran's Karaj Dam Affair: Emerging Mass Consumerism, the Politics of Promise, and the Cold War in the Third World." *Comparative Studies in Society and History* 54 (2012): 612–43.

Schlesinger, Arthur M., Jr. *The Vital Center: The Politics of Freedom.* Boston: Houghton Mifflin, 1949.

Schofield, Clive. "Elusive Security: The Military and Political Geography of South Lebanon." *GeoJournal* 31 (1993): 149–61.

Scott, James. *Seeing like a State: How Certain Schemes to Improve the Human Condition Have Failed.* New Haven, CT: Yale University Press, 1998.

Serres, Michel (with Bruno Latour). *Conversations on Science, Culture, and Time.* Translated by Roxanne Lapidus. Ann Arbor: University of Michigan Press, 1995.

Sharp, Joanne P. *Condensing the Cold War: Reader's Digest and American Identity.* Minneapolis: University of Minnesota Press, 2001.

Showers, Kate B. "Congo River's Grand Inga Hydroelectricity Scheme: Linking Environmental History, Policy and Impact." *Water History* 1 (2009): 31–58.

Six, Clemens. "The Rise of Postcolonial States as Donors: A Challenge to the Development Paradigm?" *Third World Quarterly* 30 (2009): 1103–21.

Slater, David. "Geopolitical Imaginations across the North-South Divide: Issues of Difference, Development and Power." *Political Geography* 16 (1997): 631–53.

———. *Geopolitics and the Post-colonial: Rethinking North-South Relations.* Malden, MA: Blackwell, 2004.

———. "Reimagining the Geopolitics of Development: Continuing the Dialogue." *Transactions of the Institute of British Geographers* 19 (1994): 233–38.

Smith, Neil. *American Empire: Roosevelt's Geographer and the Prelude to Globalization.* Berkeley: University of California Press, 2004.

Sneddon, Chris. "Reconfiguring Scale and Power: the Khong-Chi-Mun Project in Northeast Thailand." *Environment and Planning A* 35 (2003): 2229–50.

———. "Water, Governance and Hegemony." In *Contemporary Water Governance in the Global South: Scarcity, Marketization and Participation*, edited by L. Harris, J. Goldin, and C. Sneddon, 13–24. New York: Routledge, 2013.

Sneddon, Chris, and Coleen Fox. "Inland Capture Fisheries and Large River Systems: A Political Economy of Mekong Fisheries." *Journal of Agrarian Change* 12 (2012): 279–99.

———. "Rethinking Transboundary Waters: A Critical Hydropolitics of the Mekong Basin." *Political Geography* 25 (2006): 181–202.

———. "Water, Geopolitics, and Economic Development in the Conceptualization of a Region." *Eurasian Geography and Economics* 53 (2012): 143–60.

Spencer, John H. *Ethiopia, the Horn of Africa, and U.S. Policy.* Cambridge, MA: Institute for Foreign Policy Analysis, 1977.

Stamm, Gilbert G. "Bureau of Reclamation International Technical Assistance in Development of Arid Lands." Paper presented at the International Conference on Arid Lands in a Changing World, Tucson, Arizona, June 5, 1969.

Steinberg, Ted. *Nature Incorporated: Industrialization and the Waters of New England.* Amherst: University of Massachusetts Press, 1991.

Stine, Jeffrey, and Joel Tarr. "At the Intersection of Histories: Technology and the Environment." *Technology and Culture* 39 (1998): 601–40.

St. Louis, Vincent L., Carol A. Kelly, Éric Duchemin, John W. M. Rudd, and David M. Rosenberg. "Reservoir Surfaces as Sources of Greenhouse Gases to the Atmosphere: A Global Estimate." *BioScience* 50 (2000): 766–75.

Storey, Brit Allan. "The Bureau of Reclamation and International Water Development." Paper presented at the University of Wales Symposium, *Water in History: Global Perspectives on Politics, Economy, and Culture*, Aberystwyth, Wales, July 8–11, 1999.

Straus, Michael. *26,000 Miles along Reclamation Street: Report on South Asian Food, Water, and Power Development.* Washington, DC: Department of the Interior, 1951.

———. *Why Not Survive?* New York: Simon and Schuster, 1955.

Strout, Richard Lee. "Kilowatts for the Lamps of China." *New Republic*, March 26, 1945, 411–13.

Sun, Yi. "Militant Diplomacy: The Taiwan Strait Crises and Sino-American Relations, 1954–1958." In *The Eisenhower Administration, the Third World, and the Globalization of the Cold War*, edited by Kathryn C. Statler and Andrew L. Johns, 125–50. Lanham, MD: Rowman & Littlefield, 2006.

Swain, Ashok. "Challenges for Water Sharing in the Nile Basin: Changing Geo-politics and Changing Climate." *Hydrological Sciences Journal* 56 (2011): 687–702.

———. "Ethiopia, the Sudan, and Egypt: The Nile River Dispute." *Journal of Modern African Studies* 35 (1997): 675–94.

Swyngedouw, Eric. "Technonatural Revolutions: The Scalar Politics of Franco's Hydro-social Dream for Spain, 1939–1975." *Transactions of the Institute of British Geographers* 32 (2007): 9–28.

Thatte, C. D. "Aftermath, Overview and an Appraisal of Past Events Leading to Some of the Imbalances in the Report of the World Commission on Dams." *Water Resources Development* 17 (2001): 343–51.

Thorp, James H., Martin C. Thoms, and Michael D. Delong. "The Riverine Ecosystem Synthesis: Biocomplexity Across Space and Time." *River Research and Applications* 22 (2006): 123–47.

Tsing, Anna. *Friction: An Ethnography of Global Connections*. Princeton, NJ: Princeton University Press, 2005.

Tsou, Tang. *America's Failure in China, 1941–50*. Chicago: University of Chicago Press, 1963.

Tucker, Richard P. "Containing Communism by Impounding Rivers: American Strategic Interests and the Global Spread of High Dams in the Early Cold War." In McNeill and Unger, *Environmental Histories of the Cold War*, 139–63.

Tullos, Desiree, Bryan Tilt, and Catherine Reidy Liermann. "Introduction to the Special Issue: Understanding and Linking the Biophysical, Socioeconomic and Geopolitical Effects of Dams." *Journal of Environmental Management* 90 (2009): S203–S207.

United Nations Economic Commission for Asia and the Far East. *Development of Water Resources in the Lower Mekong Basin*. Bangkok: ECAFE, 1957.

———. *Multiple-Purpose River Basin Development*. Part 2D, *Water Resources Development in Afghanistan, Iran, Republic of Korea and Nepal*. Bangkok: ECAFE, 1961.

United Nations Economic, Social, and Cultural Organization. *Facing the Challenges: United Nations World Water Development Report 3, Case Studies Volume*. Paris: UNESCO, 2009.

United Nations Special Fund. *Report on Survey of the Awash River Basin, General Report*. Rome: FAO, 1965.

United States Agency for International Development. *To Tame a River*. Washington, DC: USAID, 1968.

Usher, Ann Danaiya. "The Race for Power in Laos: The Nordic Connections." In *Environmental Change in South-East Asia: People, Politics and Sustainable Development*, edited by Michael J. G. Parnwell and Raymond Bryant, 117–38. London: Routledge, 1996.

Vestal, Theodore M. *The Lion of Judah in the New World: Emperor Haile Selassie of Ethiopia and the Shaping of Americans' Attitudes toward Africa*. Santa Barbara, CA: Praeger, 2011.

Vörösmarty, Charles J., Michel Meybeck, Balázs Fekete, Keshav Sharma, Pamela Green, and James P. M. Syvitski. "Anthropogenic Sediment Retention: Major Global Impact from Registered River Impoundments." *Global and Planetary Change* 39 (2003): 169–90.

Vörösmarty, Charles J., Keshav P. Sharma, Balázs M. Fekete, Arthur H. Copeland, Jonathan Holden, John Marble, and John A. Lough. "The Storage and Aging of Continental Runoff in Large Reservoir Systems of the World." *Ambio* 26 (1997): 210–19.

Warne, William E. *The Bureau of Reclamation*. New York: Praeger, 1973.

———. *Mission for Peace: Point 4 in Iran*. Indianapolis: Bobbs-Merrill, 1956.

Wescoat, James. "'Watersheds' in Regional Planning." In *The American Planning Tradition: Culture and Policy*, edited by Robert Fishman, 147–72. Washington, DC: Wilson Center, Smithsonian Institution, 2000.

———. "Wittfogel East and West: Changing Perspectives on Water Development in South Asia and the US, 1670–2000." In *Cultural Encounters with the Environment: Enduring and Evolving Geographic Themes*, edited by Alexander B. Murphy and Douglas L. Johnson, 109–32. Lanham, MD: Rowman & Littlefield, 2000.

Wescoat, James, Sarah Halvorson, and Daanish Mustafa. "Water Management in the Indus Basin of Pakistan: A Half-Century Perspective." *Water Resources Development* 16 (2000): 391–406.

Wescoat, James, Roger Smith, and David Schaad. "Visits to the U.S. Bureau of Reclamation from South Asia and the Middle East, 1946–1990: An Indicator of Changing International Program and Politics." *Irrigation and Drainage Systems* (1992): 55–67.

Westad, Odd Arne. *The Global Cold War: Third World Interventions and the Making of Our Times.* Cambridge: Cambridge University Press, 2005.

Wester, Philippus, Edwin Rap, and Sergio Vargas-Velazquez. "The Hydraulic Mission and the Mexican Hydrocracy: Regulating and Reforming the Flows of Water and Power." *Water Alternatives* 2 (2009): 395–415.

White, Damian, and Chris Wilbert. "Introduction: Technonatural Time-Spaces." *Science as Culture* 15 (2006): 95–104.

White, Hayden. *The Content of the Form: Narrative Discourse and Historical Representation.* Baltimore, MD: Johns Hopkins University Press, 1987.

White, Richard. *The Organic Machine.* New York: Hill and Wang, 1995.

Winchester, Simon. *River at the Center of the World: A Journey Up the Yangtze, and Back in Chinese Time.* New York: Picador, 2004.

Winner, Langdon. "Do Artifacts Have Politics?" *Daedalus* 109 (1980): 121–36.

Wolf, Eric R. *Europe and the People without History.* Berkeley: University of California Press, 1982.

Wolman, Abel, and W. H. Lyles. *John Lucian Savage, 1879–1967, Biographical Memoir.* Washington, DC: National Academy of Sciences, 1978.

Woods, Ngaire. "Whose Aid? Whose Influence? China, Emerging Donors and the Silent Revolution in Development Assistance." *International Affairs* 84 (2008): 1205–21.

World Bank Group. *Directions in Hydropower.* Washington, DC: World Bank Group, March 2009.

World Commission on Dams (WCD). *Dams and Development: A New Framework for Decision-Making. A Report of the World Commission on Dams.* London: Earthscan, 2000.

Worster, Donald. "The Hoover Dam: A Study in Domination." In Goldsmith and Hildyard, *Social and Environmental Effects of Large Dams*, vol. 2, *Case Studies*, 17–24.

———. *Rivers of Empire: Water, Aridity, and the Growth of the American West.* New York: Pantheon Books, 1985.

Yazdanpanah, Masoud, Michael Thompson, Dariush Hayati, and Gholam Hosein Zamani. "A New Enemy at the Gate: Tackling Iran's Water Super-Crisis by Way of Transition from Government to Governance." *Progress in Development Studies* 13 (2013): 177–94.

Young, Robert J. C. *Postcolonialism: An Historical Introduction.* Oxford: Blackwell, 2001.

Zanasi, Margherita. "Exporting Development: The League of Nations and Republican China." *Comparative Studies in Society and History* 49 (2007): 143–69.

Zeitoun, Mark, Karim Eid-Sabbagh, Michael Talhami, and Muna Dajani. "Hydro-Hegemony in the Upper Jordan Waterscape: Control and Use of the Flows." *Water Alternatives* 6 (2013): 86–106.

Zewde, Bahru. *A History of Modern Ethiopia, 1855–1991.* 2nd ed. Athens: Ohio University, 2001.